OSHA and EPA PROCESS SAFETY MANAGEMENT REQUIREMENTS

A Practical Guide for Compliance

Mark S. Dennison
Attorney at Law

JOHN WILEY & SONS, INC.
New York • Chichester • Weinheim • Brisbane • Singapore • Toronto

This text is printed on acid-free paper.

Copyright © 1994 by John Wiley & Sons, Inc.

All rights reserved. Published simultaneously in Canada.

No part of this publication may be reproduced, stored in a retrieval system or transmitted in any form or by any means, electronic, mechanical, photocopying, recording, scanning or otherwise, except as permitted under Sections 107 or 108 of the 1976 United States Copyright Act, without either the prior written permission of the Publisher, or authorization through payment of the appropriate per-copy fee to the Copyright Clearance Center, 222 Rosewood Drive, Danvers, MA 01923, (978) 750-8400, fax (978) 750-4470. Requests to the Publisher for permission should be addressed to the Permissions Department, John Wiley & Sons, Inc., 111 River Street, Hoboken, NJ 07030, (201) 748-6011, fax (201) 748-6008.

Printed in the United States of America

Library of Congress Cataloging-in-Publication Data

Dennison, Mark S.
    OSHA and EPA process safety management requirements: a practical guide for compliance / Mark S. Dennison.
    p. cm.
    Includes index.
    ISBN 0-471-28641-9
    1. Industrial Safety—United States. 2. Industrial Safety—Law and legislation—United States. 3. Chemical plants—United States—safety measures. I. Title.
T55.D46 1994
363.17'6—dc20   94-12507
                                                                         CIP

10 9 8 7 6 5 4 3

To Tracey, who provides a steady flame for my candle in the wind
AND
To the kids, Erin, Johnny, Jessie, Ginny, and Timmy, for their smiles and their laughter

Contents

About the Author xiii

Acknowledgements xiv

Preface xv

PART I - OSHA'S PSM STANDARD

Chapter 1
Background on OSHA's Process Safety Management Standard *1*

- 1.1 CHEMICAL INCIDENTS AND PROCESS SAFETY MANAGEMENT (PSM) EFFORTS 1
  - 1.1.1 International Efforts Concerning Process Safety 2
  - 1.1.2 U.S. Regulation of Hazardous Chemicals 2
  - 1.1.3 Other OSHA Safety Standards 3
  - 1.1.4 Industry Efforts Concerning Process Safety Management 4
- 1.2 DEVELOPMENT OF THE PSM STANDARD 5
  - 1.2.1 Impact of Clean Air Act Amendments of 1990 6
- 1.3 PROVISIONS OF THE PSM STANDARD 8
- 1.4 INDUSTRIES AFFECTED BY THE PSM STANDARD 9
  - 1.4.1 Concerns Raised During the Rulemaking Process 9
  - 1.4.2 Concerns Raised by the Office of Management and Budget 11
  - 1.4.3 Consideration of Other Regulatory Options 11
- 1.5 LIST OF REGULATED CHEMICALS AND THRESHOLD QUANTITIES 12
- 1.6 EFFECTIVENESS RATE OF PSM STANDARD 13
  - 1.6.1 Cost Impacts of the PSM Standard 13
  - 1.6.2 Impact on Small Businesses 14
  - 1.6.3 Sunset Provision Unnecessary 15

vi Contents

Chapter 2
Process Safety Management Regulatory Framework 17

2.1 OVERVIEW OF THE PSM STANDARD 17
    2.1.1 Impetus Behind the New PSM Standard 18
    2.1.2 Performance-Oriented Approach 18
2.2 BASIC PSM STANDARD REQUIREMENTS 19
2.3 APPLICATION OF THE PSM STANDARD: PARAGRAPH (a) 20
    2.3.1 Meaning of the Term "Processes" 21
    2.3.2 List of Highly Hazardous Chemicals 22
    2.3.3 Minimum Threshold Levels 23
    2.3.4 Role of EPCRA and CAAA Chemical Lists 24
    2.3.5 Exemptions for Hydrocarbon Fuels 25
    2.3.6 Flammable Liquids Exemption 25
    2.3.7 Other Exemptions 26
2.4 DEFINITIONS: PARAGRAPH (b) 28
2.5 HOT WORK PERMIT: PARAGRAPH (k) 30
    2.5.1 Hot Work Permit Compliance Checklist 31
2.6 TRADE SECRETS: PARAGRAPH (p) 38
    2.6.1 Trade Secrets Compliance Checklist 41
2.7 STATE PROGRAMS 43
    2.7.1 Delaware 43
    2.7.2 California 45
2.8 OSHA COMPLIANCE DIRECTIVE 46
2.9 OSHA INSPECTIONS AND ENFORCEMENT 46
    2.9.1 Inspection Totals 47
    2.9.2 PSM Provisions Cited 48
    2.9.3 Violations of Operating Procedures 48
    2.9.4 Specific Violations 49
    2.9.5 $1.6 Million Fine Proposed for Violation of PSM Standard 49
    2.9.6 Specific Citations 50
    2.9.7 Other Penalties 51
    2.9.8 Settlement of 1989 Phillips Petroleum Explosion 51

Chapter 3
PSM Program Implementation and Operation 53

3.1 COMPONENTS OF A PSM SYSTEM 53
    3.1.1 Generic Characteristics of a Management System 54

            3.1.2   Components of a Process Safety Management
                    Program 55
            3.1.3   Compiling Process Safety Information 59
3.2   PROCESS SAFETY INFORMATION: SPECIFIC
       REQUIREMENTS 59
            3.2.1   Process Safety Information: Compliance Tips 62
            3.2.2   Process Safety Information: Compliance Checklist
                    65
3.3   EMPLOYEE INVOLVEMENT IN PROCESS SAFETY
       MANAGEMENT 69
3.4   EMPLOYEE PARTICIPATION: SPECIFIC
       REQUIREMENTS 69
            3.4.1   Employee Participation Compliance Checklist 71
3.5   PROCESS SAFETY OPERATING PROCEDURES 73
            3.5.1   Written Operating Procedures 73
3.6   OPERATING PROCEDURES: SPECIFIC
       REQUIREMENTS 74
            3.6.1   Operating Procedures and Practices 77
            3.6.2   Operating Procedures Compliance Checklist 79

Chapter 4
Conducting the Process Hazard Analysis *83*

4.1   PHA METHODOLOGIES 83
4.2   SELECTING THE APPROPRIATE METHODOLOGY 86
4.3   PHA BENEFITS FOR SMALL BUSINESSES 87
4.4   PROCESS HAZARD ANALYSIS: SPECIFIC
       REQUIREMENTS 88
4.5   TEAM APPROACH 91
4.6   PROCESS HAZARD ANALYSIS: COMPLIANCE
       GUIDELINES 93
            4.6.1   Process Hazard Analysis Compliance Checklist 94
            4.6.2   PHA Team 101
            4.6.3   Application of PHA to Specific Processes 101
            4.6.4   Small Businesses 102

Chapter 5
Training Employees and Using Contractors *104*

5.1   EMPLOYEE TRAINING PROGRAMS 104

viii Contents

5.2 TRAINING: SPECIFIC REQUIREMENTS 105
    5.2.1 Application 106
    5.2.2 Grandfathering 106
    5.2.3 Refresher Training 107
    5.2.4 Training Documentation 107
    5.2.5 Clean Air Act Training Requirements 108
5.3 EMPLOYEE TRAINING COMPLIANCE GUIDELINES 109
    5.3.1 Training Examples 109
    5.3.2 Training Program Evaluation 110
    5.3.3 Training Compliance Checklist 111
5.4 APPLICATION OF PSM STANDARD TO CONTRACTORS 114
    5.4.1 Use of Contractors 115
    5.4.2 Employer Responsibilities 115
    5.4.3 Contract Employer Responsibilities 116
5.5 CONTRACTORS: SPECIFIC PSM REQUIREMENTS 116
    5.5.1 Contractors: Compliance Guidelines 121
    5.5.2 Contractors Compliance Checklist 123
    5.5.3 Nonroutine Work Authorizations 128
    5.5.4 Contractor Training Settlement 128

Chapter 6
Maintaining Mechanical Integrity and Managing Process Changes *130*

6.1 MECHANICAL INTEGRITY OF EQUIPMENT 130
6.2 MECHANICAL INTEGRITY: SPECIFIC REQUIREMENTS 131
    6.2.1 Mechanical Integrity: Compliance Guidelines 134
    6.2.2 Mechanical Integrity Compliance Checklist 136
    6.2.3 Lines of Defense 142
    6.2.4 Inspection Methodologies 143
    6.2.5 Quality Assurance System 144
6.3 MANAGEMENT OF PROCESS CHANGES 145
    6.3.1 Developing a Management of Change System 145
6.4 MANAGEMENT OF CHANGE: PSM STANDARD REQUIREMENTS 147
    6.4.1 Managing Change: Compliance Guidelines 148
    6.4.2 Management of Change Compliance Checklist 150

Contents ix

        6.4.3   Request for Change Form   153
        6.4.4   Sample Request for Change Form   154
6.5   PRE-STARTUP SAFETY REVIEWS   155
        6.5.1   Pre-startup Safety Review: PSM Standard Requirements   155
        6.5.2   Pre-Startup Safety: Compliance Guidelines   157
        6.5.3   Pre-Startup Safety Review Compliance Checklist   158

Chapter 7
Incident Investigation, Emergency Planning, and Compliance Audits *161*

7.1   INVESTIGATION AND RESPONSE TO PROCESS INCIDENTS   161
        7.1.1   Case Histories of Process Incidents   161
7.2   INCIDENT INVESTIGATION   164
        7.2.1   Incident Investigation: Specific Requirements   165
        7.2.2   Investigation of Incidents: Compliance Guidelines   167
        7.2.3   Incident Investigations Compliance Checklist   171
7.3   EMERGENCY PLANNING AND RESPONSE: SPECIFIC REQUIREMENTS   174
        7.3.1   Emergency Preparedness: Compliance Guidelines   175
        7.3.2   Emergency Planning and Response Compliance Checklist   178
        7.3.3   Pre-Planning for Releases   184
        7.3.4   Plant and Local Community Coordination of Emergency Response   184
7.4   PSM COMPLIANCE AUDITS   185
        7.4.1   Compliance Audits: Specific Requirements   185
        7.4.2   Compliance Audits: PSM Standard Compliance Guidelines   186
        7.4.3   Compliance Audits: Compliance Checklist   188
        7.4.4   Audit Team   190
        7.4.5   Review of Relevant Documentation   190
        7.4.6   Corrective Action   191
        7.4.7   Preparing for the Audit   192
        7.4.8   Checklist of Items to Examine During the Audit   192

x Contents

PART II - EPA'S CHEMICAL ACCIDENT RELEASE
PREVENTION REGULATIONS

Chapter 8
Background on EPA's Chemical Accident Release Prevention
Regulations *197*

8.1 THE NEW EPA RULE 197
8.2 COMPARISON OF OSHA AND EPA RULES 198
8.3 ACCIDENTAL RELEASE SCENARIOS 199
8.4 IMPETUS BEHIND THE EPA RULE 200
8.5 EPA'S ACCIDENTAL RELEASE INFORMATION
 PROGRAM 201
8.6 CLEAN AIR ACT AMENDMENTS OF 1990 202
8.7 RISK MANAGEMENT PROGRAMS: CLEAN AIR ACT
 REQUIREMENTS 203
8.8 COST ESTIMATES 206

Chapter 9
Risk Management Plans *208*

9.1 PURPOSE OF EPA'S NEW RULE 208
    9.1.1 Applicability of the EPA Rule 209
    9.1.2 Important Definitions 210
    9.1.3 Regulated Substances and Thresholds 210
    9.1.4 Risk Management Program Elements 211
9.2 HAZARD ASSESSMENT 211
    9.2.1 Worst-Case Release Scenarios 211
    9.2.2 Other Accidental Release Scenarios 212
    9.2.3 Potential Off-site Consequences 213
    9.2.4 Use of Generic Methodologies for Assessing
          Off-site Impacts 214
    9.2.5 Range of Events 215
    9.2.6 Updating of Off-site Consequences Analyses 215
    9.2.7 Five-Year History of Releases 216
9.3 PREVENTION PROGRAM 216
    9.3.1 Development of a Management System 218
    9.3.2 Process Hazard Analysis 218
    9.3.3 Application of PHA Techniques 220
    9.3.4 Monitoring and Detection Systems 221

Contents xi

        9.3.5   Compiling Process Safety Information 222
        9.3.6   Standard Operating Procedures (SOPs) 223
        9.3.7   Training Requirements 224
        9.3.8   Maintenance (Mechanical Integrity) 225
        9.3.9   Pre-Startup Review 226
        9.3.10 Management of Change 227
        9.3.11 Safety Audits 228
        9.3.12 Accident Investigation 229
        9.3.13 Evaluation of Root-Causes and Near-Misses 229
9.4   EMERGENCY RESPONSE 230
9.5   RISK MANAGEMENT PLAN AND DOCUMENTATION 232
        9.5.1   RMP Content 233
        9.5.2   Submission of RMPs 235
        9.5.3   Auditing of RMPs 236
        9.5.4   Registration: Information Required 237
        9.5.5   Implementation of Registration Requirements 239
9.6   PENALTIES FOR VIOLATIONS OF THE RULE 240
9.7   SECTION BY SECTION REVIEW OF EPA'S RULE 240

Chapter 10
Comparison Between EPA's CARP Regulations and OSHA's PSM Standard *247*

10.1   RELATIONSHIP BETWEEN EPA'S RULE AND OSHA'S PSM STANDARD 247
10.2   DIFFERENCES BETWEEN EPA'S RULE AND OSHA'S PSM STANDARD 248
10.3   SECTION-BY-SECTION COMPARISON OF EPA'S CARP REGULATIONS AND OSHA'S PSM STANDARD 250
        10.3.1 Use of Contractors 253
        10.3.2 Compliance Dates 254
10.4   ROLE OF STATE LAWS 254
10.5   OTHER APPROACHES CONSIDERED 255

APPENDIXES *259*

Appendix 1
PSM Standard (29 CFR 1910.119) *261*

Appendix 2
OSHA List of Highly Hazardous Chemicals, Toxics, and Reactives *275*

Appendix 3
Block Flow Diagram and Simplified Process Flow Diagram *279*

Appendix 4
Sources of Further Information on Process Safety Management *281*

Appendix 5
Directory of Consultation Programs *284*

Appendix 6
States With Approved OSHA Programs *286*

Appendix 7
EPA List of Regulated Substances and Thresholds for Accidental Release Prevention *289*

Appendix 8
Accidental Release Prevention Provisions (Proposed) *307*

Appendix 9
OSHA Compliance Directive CPL 2-2.45A *330*

REFERENCES *387*

ACRONYMS AND ABBREVIATIONS *389*

INDEX *391*

About the Author

Mark Dennison is an attorney and author or co-author of numerous books and articles on environmental and land use law issues. His many books include *Understanding Solid and Hazardous Waste Identification and Classification* (1993); *Hazardous Waste Regulation Handbook: A Practical Guide to RCRA and Superfund* (1994); *Wetlands and Coastal Zone Regulation and Compliance*, with Steven Silverberg (1993); and *Wetlands: Science, Law, and Technology*, with James Berry (1993). Mr. Dennison is also the editor-in-chief of the monthly newsletter, *Environmental Strategies for Real Estate*, and serves on the editorial advisory boards of *Occupational Health & Safety* magazine and *Environmental Protection* magazine. He is in private practice in New Jersey, where he specializes in environmental, land use, real estate and zoning law. He is admitted to practice in New Jersey and New York. Mr. Dennison holds a B.A., *magna cum laude*, from the State University of New York (Oswego), a M.A. from Syracuse University, and a J.D. from New York Law School. He is also working toward completion of an LL.M. degree in environmental law at Pace University School of Law.

Acknowledgments

I wish to express my appreciation to everyone who played a role in publication of this book. I thank the editorial and production staff at Van Nostrand Reinhold, including Ken McCombs, Mike Sherry, and Barbara Mathieu, for all their assistance. And, I especially wish to thank Alex Padro, Acquisitions Editor, for motivating me to write this book, providing valuable editorial guidance, and most importantly, for adding such good humor to an otherwise painstaking process.

I also wish to thank the many peer reviewers who provided helpful suggestions for improving the organization, focus, and coverage of the book. And, I must say thank you to the many helpful individuals working at different federal and state regulatory offices who answered my questions and provided copies of various forms and government documents.

Additional thanks go to family and friends for providing moral support and inspiration, especially Erin Carrather, Jonathan Huff, Bobbie Waits, Harry ("The Grip") Huff, Mrs. J. Howard Gould, Keith Dennison, Randy Masters, and Jim Berry.

Most of all, wholehearted thanks must go to my greatest source of inspiration, Tracey Huff, without whose constant love and support this book could never have been written. Thank you my love.

M.S.D.

Preface

In 1992, in response to the accidental release provision of the Clean Air Act Section 112(r), OSHA issued its process safety management (PSM) regulations. OSHA's PSM standard mainly applies to manufacturing industries, particularly those involved with chemicals, transportation equipment, and fabricated metal products. In each industry, the PSM standard applies to those companies with processes involving any of more than 130 listed toxic and reactive chemicals above threshold quantities.

OSHA saw a pressing need for the new PSM Standard in light of the ongoing occurrence of serious accidents at chemical processing plants, which emphasized the need for tighter regulatory control over hazards associated with processes that use or produce highly hazardous chemicals.

Human error and inadequate management systems were identified as the most significant contributing factors to major plant accidents. Recognizing that human error can be minimized, and better management systems can be implemented to prevent such disasters, OSHA's new PSM Standard sets forth a performance-based approach to managing hazards related to use and production of highly hazardous chemicals.

In addition to OSHA's introduction of the PSM standard in 1992, EPA followed in 1993 with issuance of proposed regulations governing Risk Management Plans (RMPs) for Chemical Accident Release Prevention. These regulations should be finalized near the end of 1994. The EPA program, also initiated in response to the Clean Air Act's accidental release provision, will be much broader in scope, and is expected to impact more than 118,000 facilities, many of which are outside the realm of the "traditional" chemical manufacturing industry.

OSHA and EPA Process Safety Management Requirements: A Practical Guide for Compliance describes the various requirements of OSHA's new Process Safety Management (PSM) Standard and EPA's new regulations on Risk Management Plans for Chemical

Accidental Release Prevention. Using easy-to-understand language, the book takes the reader step-by-step through each regulatory requirement and provides "how-to" compliance advice for process safety managers, safety engineers and compliance officers, safety consultants, and attorneys who advise the chemical industry, including:

> *"Nuts and bolts" explanation of OSHA's PSM Standard and EPA's CARP regulations
> *Specific compliance measures for each regulatory requirement
> *Helpful Process Hazard Analysis methodologies
> *Case studies illustrating ways to prevent plant accidents
> *Guidelines for developing a management of change programs
> *Performance of PSM compliance audits
> *Criteria for conducting pre-startup safety reviews and ensuring mechanical integrity of process safety equipment
> *Using employee participation to maintain process safety
> *Importance of incident investigation and emergency preparedness
> *Helpful comparison of the OSHA and EPA programs
> *Practical steps for implementing EPA Risk Management Plans.
> *Useful discussion of government inspections and penalties for noncompliance

This is the first and only book to cover EPA's new regulations on Risk Management Programs for Chemical Accident Release Prevention. The book offers specific guidance on each component of the EPA-required risk management programs and covers the regulated substances and their threshold quantities for accidental release prevention.

The book contains helpful compliance checklists, as well as useful appendices, including lists of regulated substances and threshold quantities, important government contacts, and OSHA's PSM Compliance Directive CPL 2-2.45A.

Mark S. Dennison

PART I - OSHA'S PSM STANDARD

Chapter 1

Background on OSHA's Process Safety Management Standard

1.1 CHEMICAL INCIDENTS AND PROCESS SAFETY MANAGEMENT (PSM) EFFORTS

Releases of toxic, reactive or flammable liquids and gases in processes involving highly hazardous chemicals have been reported for many years. Incidents continue to occur in a variety of industries that use a variety of highly hazardous chemicals which may be toxic, reactive, flammable, or explosive or exhibit a combination of these attributes. Regardless of the industry that uses these highly hazardous chemicals, there exists a potential for an accidental release if a highly hazardous chemical is not properly controlled. This in turn presents the potential for a devastating incident. Recent major incidents include the 1984 Bhopal incident resulting in more than 2,000 deaths; the October 1989 Phillips 66 plastics plant incident resulting in 23 deaths and some 350 injuries; the July 1990 Arco Chemical Company incident resulting in 17 deaths; the July 1990 BASF incident resulting in 2 deaths and 41 injuries; and the May 1991 IMC incident resulting in 8 deaths and 128 injuries.

While these major incidents involving highly hazardous chemicals have drawn national attention to the potential for major catastrophes, many other releases of highly hazardous chemicals have occurred in recent years. These releases continue to pose a significant threat to workers at facilities that use, manufacture, and handle extremely hazardous chemicals. The continuing occurrence of incidents has provided impetus, internationally and nationally, for authorities to develop or consider the development of legislation and regulations directed toward eliminating or minimizing the potential for such events.

2   OSHA and EPA PSM Requirements

1.1.1   International Efforts Concerning Process Safety

International efforts have included the development of the Seveso Directive by the European Economic Community after several large scale incidents occurred in the 1970s, including the disasters that occurred in Flixborough, England and Seveso, Italy. This Directive addresses the major accident hazards of certain industrial activities, lists the hazardous materials of concern and is directed toward controlling those activities that could give rise to major accidents in an effort to protect the environment and human safety and health. Subsequent international efforts have included the development of guidelines for identifying, analyzing and controlling major hazard installations in developing countries and a hazards assessment manual which provides measures to control major hazard accidents developed by the World Bank; the development of the Code of Practice on the Prevention of Major Accident Hazards by the International Labour Organization; and the special conferences held by the Organization of Economic and Cooperative Development to consider the catastrophic potential of accidents involving hazardous substances and the means to prevent their occurrence and mitigate their impact.

1.1.2   U.S. Regulation of Hazardous Chemicals

In the United States, Congress, federal agencies, state governments, industry, unions and other interested groups have become actively concerned and involved with protecting employees, the public, and the environment from major chemical accidents involving highly hazardous chemicals. In 1985, in response to the potential for catastrophic releases, the Environmental Protection Agency (EPA) initiated a program to encourage community planning and preparation relative to serious hazardous materials releases. In 1986, Congress passed the framework for emergency planning efforts through Title III of the Superfund Amendments and Reauthorization Act (SARA), also known as the Emergency Planning and Community Right-to-Know Act (42 U.S.C. 11001 *et seq.*). SARA encourages and supports states and local communities in efforts to address the problems of chemical releases. Under Section 302 of SARA, 42 U.S.C. 11002, EPA was required to publish a list of extremely

Chapter 1 - Background on PSM Standard 3

hazardous substances with threshold planning quantities which would trigger planning in states and local communities (52 Fed. Reg. 13378).

After the 1984 Bhopal, India incident involving an accidental release of methyl isocyanate which resulted in more than 2,000 deaths, the Occupational Safety and Health Administration (OSHA) determined that it was necessary to immediately investigate U.S. producers and users of methyl isocyanate. This investigation indicated that while the chemical industry is subject to OSHA's general industry standards, these standards do not presently contain specific coverage for chemical industry process hazards, nor do they specifically address employee protection from large releases of hazardous chemicals.

1.1.3 Other OSHA Safety Standards

OSHA standards do exist for employee exposure to certain specific toxic substances (see subpart Z of part 1910), and hazardous chemicals are covered generally by other OSHA standards such as the Hazard Communication Standard, 29 CFR Section 1910.1200. While these standards do address hazardous chemicals, they focus on routine or daily exposures. In many cases, they also address emergencies such as spills, however, they do not address the precautions necessary to prevent large accidental releases that could result in catastrophes.

Additionally, OSHA has certain standards contained in subpart H of 29 CFR part 1910, Hazardous Materials, concerning flammable liquids, compressed and liquified petroleum gases, explosives and fireworks. The flammable liquids and compressed and liquified petroleum gas standards emphasize equipment specification and the flammability of materials and do not thoroughly address other hazards of materials such as toxicity, and the standard concerning explosives and fireworks does not address the hazards involved during their manufacture. Beyond these standards, OSHA must depend on section 5(a)(1) of the Occupational Safety and Health Act, the general duty clause, to protect employees from other hazardous situations arising from the use of highly hazardous chemicals in certain industrial processes and must use national consensus standards and industry standards to support these general duty clause citations.

4   OSHA and EPA PSM Requirements

1.1.4   Industry Efforts Concerning Process Safety Management

Industry has taken measures aimed at improving the protection of public health and safety by improving chemical process safety to prevent releases. The Chemical Manufacturers Association (CMA) developed the Chemical Awareness and Emergency Response Program to foster cooperation, knowledge and response within communities. Additionally CMA produced a report on process safety management, "Process Safety Management, (Control of Acute Hazards)," in order to increase knowledge among CMA members about systematic approaches to process safety analysis.

In 1985, the American Institute of Chemical Engineers, formed a separate branch, the Center for Chemical Process Safety (the Center). The Center's charter is to develop and disseminate technical information to be used in the prevention of major chemical accidents. The Center has become well known for its process safety management guidance publications. Also an industry consulting group, the Organization Resources Counselors (ORC), and an industry trade association, the American Petroleum Institute (API), have developed recommended practices to address the protection of employees and the public through the prevention or mitigation of the effects of dangerous chemical releases. In 1990 API published its Recommended Practice 750, Management of Process Hazards, "to provide a more structured and formal approach to existing practices and to ensure a comprehensive approach to process safety."

Unions representing employees immediately exposed to danger from processes using highly hazardous chemicals have demonstrated a great deal of interest and activity in controlling major chemical accidents. For example, the International Confederation of Free Trade Unions and the International Federation of Chemical, Energy and General Workers' Unions issued a special report on the Bhopal, India accident. Additionally the United Steelworkers of America investigated and issued a special report on the 1988 PEPCON plant oxidizer accident in Henderson, Nevada (ammonium perchlorate explosion, two deaths and 350 injuries). Further, unions including the United Steelworkers of America, the International Chemical Workers, and the Oil, Chemical and Atomic Workers, have undertaken large-scale efforts to train and educate their members who work in the petrochemical industry.

1.2 DEVELOPMENT OF THE PSM STANDARD

The need to focus on safety and health in the chemical industry was reinforced in August 1985. A serious release of highly hazardous chemicals (aldicarb oxime and methyl chloride) occurred at a plant in Institute, West Virginia. While no deaths occurred, 135 persons were injured. The experience of investigating this release indicated to OSHA that there was a need to look beyond existing standards and led OSHA to develop a demonstration program of special inspections in a small segment of the chemical industry. The purpose of the program was to examine industry practices for the prevention of disastrous releases and the mitigation of the effects of releases that do occur, and to consider ways in which OSHA could best protect employees in the industry from these hazards. Based on the results of the program, OSHA determined that chemical plant inspections need a comprehensive inspection approach, which includes physical conditions and management systems. Since this program was initiated, OSHA has issued a series of inspection directives, updated by growing experience and knowledge, that address system safety evaluations of operations with catastrophic potential. One important change in the successive directives was the expansion of the scope of facilities to be inspected. Inspections were to be conducted in industries beyond chemical manufacturing because potentially hazardous chemical releases are not limited to chemical manufacturing and similar precautions should be implemented in operations in which hazardous chemicals are used, mixed, stored or otherwise handled.

OSHA believed that available evidence supported the need for a standard and that adequate data and information existed upon which a standard could be based. Accordingly, on July 17, 1990, OSHA published in the Federal Register (55 Fed. Reg. 29150) a proposed standard containing requirements for the management of hazards associated with processes using highly hazardous chemicals in order to help assure that workers have a safe and healthful workplace.

OSHA's proposed rule emphasized the management of hazards associated with highly hazardous chemicals. The application of management controls to processes involving highly hazardous chemicals was recommended to OSHA by the Organization Resources Counselors (ORC). ORC observed: "[W]hen

6 OSHA and EPA PSM Requirements

OSHA issued its final report on the Special Emphasis Program for the Chemical Industry (Chem SEP), among its findings were that 'specification standards ... will not ... ensure safety in the chemical industry ... (because such standards) tend to freeze technology and may minimize rather than maximize employers safety efforts.' The Chem SEP report recommended a new approach to the identification and prevention of potentially catastrophic situations. This approach would involve "performance-oriented standards ... to address the overall management of chemical production and handling systems."

Further regarding the recommended standard, ORC noted that: "The recommendations it contains are a systematic approach to chemical process hazards management which, when implemented, will ensure that the means for preventing catastrophic release, fire and explosion are understood, and that the necessary preventive measures and lines of defense are installed and maintained." The application of mangement controls to processes involving highly hazardous chemicals was also supported by other interested groups.

1.2.1 Impact of Clean Air Act Amendments of 1990

Approximately four months after the publication of OSHA's proposed standard for process safety management of highly hazardous chemicals, the Clean Air Act Amendments (CAAA) were enacted into law (Pub. L. 101-549, November 15, 1990). The CAAA require in section 304 that the Secretary of Labor, in coordination with the Administrator of the Environmental Protection Agency, promulgate, pursuant to the Occupational Safety and Health Act of 1970, a chemical process safety standard to prevent accidental releases of chemicals which could pose a threat to employees. The CAAA require that the standard include the development of a list of highly hazardous chemicals which include toxic, flammable, highly reactive and explosive substances. The CAAA specified the areas which must be covered by the standard. The OSHA standard must require employers to:

> (1) Develop and maintain written safety information identifying workplace chemical and process hazards, equipment used in the processes, and technology used in the processes;

(2) Perform a workplace hazard assessment, including, as appropriate, identification of potential sources of accidental releases, an identification of any previous release within the facility which had a likely potential for catastrophic consequences in the workplace, estimation of workplace effects of a range of releases, estimation of the health and safety effects of such range on employees;
(3) Consult with employees and their representatives on the development and conduct of hazard assessments and the development of chemical accident prevention plans and provide access to these and other records required under the standard;
(4) Establish a system to respond to the workplace hazard assessment findings, which shall address prevention, mitigation, and emergency responses;
(5) Periodically review the workplace hazard assessment and response system;
(6) Develop and implement written operating procedures for the chemical process including procedures for each operating phase, operating limitations, and safety and health considerations;
(7) Provide written safety and operating information to employees and train employees in operating procedures, emphasizing hazards and safe practices;
(8) Ensure contractors and contract employees are provided appropriate information and training;
(9) Train and educate employees and contractors in emergency response in a manner as comprehensive and effective as that required by the regulation promulgated pursuant to section 126(d) of the Superfund Amendments and Reauthorization Act;
(10) Establish a quality assurance program to ensure that initial process related equipment, maintenance materials, and spare parts are fabricated and installed consistent with design specifications;
(11) Establish maintenance systems for critical process related equipment including written procedures, employee training, appropriate inspections, and testing of such equipment to ensure ongoing mechanical integrity;
(12) Conduct pre-startup safety reviews of all newly installed or modified equipment;

8 OSHA and EPA PSM Requirements

(13) Establish and implement written procedures to manage change to process chemicals, technology, equipment and facilities; and

(14) Investigate every incident which results in or could have resulted in a major accident in the workplace, with any findings to be reviewed by operating personnel and modifications made if appropriate.

Also under the CAAA, the Environmental Protection Agency has specified duties relative to the prevention of accidental releases (see section 301(r)). Generally EPA is required to develop a list of chemicals and a Risk Management Plan.

1.3 PROVISIONS OF THE PSM STANDARD

The major objective of process safety management of highly hazardous chemicals is to prevent unwanted releases of hazardous chemicals especially into locations which could expose employees and others to serious hazards. An effective process safety management program requires a systematic approach to evaluating the whole process. Using this approach the process design, process technology, operational and maintenance activities and procedures, nonroutine activities and procedures, emergency preparedness plans and procedures, training programs, and other elements which impact the process are all considered in the evaluation. The various lines of defense that have been incorporated into the design and operation of the process to prevent or mitigate the release of hazardous chemicals need to be evaluated and strengthened to assure their effectiveness at each level. Process safety management is the proactive identification, evaluation and mitigation or prevention of chemical releases that could occur as a result of failures in process, procedures or equipment.

The PSM Standard targets highly hazardous chemicals that have the potential to cause a catastrophic incident. This standard as a whole is to aid employers in their efforts to prevent or mitigate episodic chemical releases that could lead to a catastrophe in the workplace and possibly to the surrounding community. To control these types of hazards, employers need to develop the necessary expertise, experiences, judgement and proactive initiative within their workforce to properly implement and maintain an effective process safety management program as envisioned in the OSHA standard.

1.4 INDUSTRIES AFFECTED BY THE PSM STANDARD

Based on a report prepared by Kearney/Centaur and a follow-up review of national chemical databases, OSHA determined that 24,939 establishments in 127 industry subgroups will be affected by the PSM Standard. The population at risk is an estimated 3.0 million workers (2.37 million plant employees and 653,000 contract employees) and is found throughout manufacturing, particularly in Standard Industrial Classification (SIC) code 28, Chemicals and Allied Products, SIC 37, Transportation Equipment, and SIC 34, Fabricated Metal Products, Except Machinery and Transportation Equipment. In addition to manufacturing, other industries have workers who are at risk, including the natural gas liquids (SIC 1321), farm product warehousing (SIC 4221), electric, gas, and sanitary services (SIC 49) and wholesale trade (SICs 50 and 51) industries. The extent of the impact will vary by industry depending on current practice, the number of processes, and the quantities of highly hazardous materials on site.

OSHA excluded from this final impact analysis establishments in California, Delaware and New Jersey, where process safety management statutes have already been enacted. In these three states the compliance burden is unaffected by the federal rule. OSHA estimated current practices with the provisions of the process safety management rule using OSHA survey data, survey data compiled by a major chemical engineering magazine, and data in the rulemaking record. For all industries affected by the proposed rule, none are currently in full compliance, although compliance is greater than 75 percent among some establishments for some specific provisions. Generally, larger firms have a higher current compliance rate than smaller firms, but for many industries the compliance rate differences by establishment size are not substantial.

1.4.1 Concerns Raised During the Rulemaking Process

Several issues were brought up during the rulemaking. First, many participants asserted that OSHA should permit required information to be stored electronically or on computers.

10 OSHA and EPA PSM Requirements

Electronic storage or computerized storage of records and information required by this standard is permissible, as long as it is readily accessible and easily understood.

Second, in Issue 10 of the proposed rule (55 Fed. Reg. 29159) OSHA asked whether provisions should be delayed or phased-in (timeframes for conducting process hazard analyses were discussed in a separate issue, Issue 3 at 29158). Participants suggested a variety of schedules. However, OSHA decided that, with the exception of allowing a phase-in period for paragraph (d), process safety information, and paragraph (e), process hazards analysis, no other phase-in period was necessary or warranted.

Third, also in Issue 10, OSHA asked whether it is necessary for all of the covered industries to meet all of the proposed provisions. OSHA was concerned about the potential impact on small businesses. Most of the participants who addressed this issue believed that small facilities should not be exempted if they have the threshold quantity of chemicals in their processes since the potential for a catastrophe is based on the amount of chemical present rather than on the size of facility. Several of these participants suggested that OSHA provide special assistance to small business employers. OSHA agreed with participants that plants should be covered based on whether they have the threshold quantity of a covered highly hazardous chemical. OSHA also agreed with the recommendation suggesting that OSHA provide special assistance to small businesses and is considering this issue at the present time. The agency is developing a compliance assistance "outreach" program to assist small businesses.

Finally, in Issue 11 (55 Fed. Reg. 29159) of the proposed rule, OSHA asked whether employers, when they have a threshold quantity of a highly hazardous chemical as specified by the standard, should be required to notify the OSHA area office of their location. Other entities which regulate potentially catastrophic workplaces require notification of the regulating authority. Numerous participants addressed this issue. Some participants believed that notification should be required. Other participants indicated that while they did not see a benefit to notification, they would not object if a notification requirement was kept simple. Still others objected to notification as an unnecessary burden, and in some cases observed that EPA already requires notification or that perhaps OSHA should access the information already required to be submitted to EPA.

For example, Organization Resources Counselors indicated that: "Before OSHA inserts such a requirement into the standard, it should determine what use it will make of such notification and whether or not the information is already available from other resources." OSHA decided not to require notification in the final standard. OSHA believed that requiring such information would be redundant with requirements that already exist under SARA and under the Clean Air Act Amendments, which contain reporting requirements. Since similar information is already required to be reported, OSHA will work with EPA to obtain needed plant location information, instead of placing a redundant burden on employers.

1.4.2 Concerns Raised by the Office of Management and Budget

On September 19, 1990, the Office of Management and Budget (OMB) filed comments on the process safety management proposal. OMB raised several concerns about the proposal. OMB observed that:

(1) OSHA failed to consider alternative regulatory options;
(2) The effectiveness of OSHA's approach is uncertain;
(3) The costs of the standard may be higher than estimated and may adversely affect profitability;
(4) The standard may have high costs and few benefits for small business employers and, therefore, could be anticompetitive; and
(5) OSHA should consider a sunset provision in the final rule that would cause the rule to expire after five years if it does not have the intended effect of providing significant reductions in the number of workplace accidents associated with hazardous chemicals.

OSHA carefully evaluated the OMB's comments and determined that the modifications to the proposed rule were responsive to the OMB concerns. OSHA's response to the major concerns are summarized in the remainder of this chapter.

1.4.3 Consideration of Other Regulatory Options

OMB stated that OSHA had failed to consider alternative regulatory options that might protect workers equally well at lower cost. OSHA's latitude to consider regulatory options, such as those contemplated by OMB, is somewhat limited by the Clean Air Act Amendments (CAAA). For example, in section 304 of the CAAA, OSHA was directed to enact a chemical process safety standard containing certain minimum elements within one year. The Clean Air Act Amendments specified 14 elements which OSHA must include in the process safety standard. OSHA included these elements in its final process safety management standard. OMB suggested that OSHA consider an alternative regulatory approach that allows firms to use the results of the hazard analysis to determine which of the other safety requirements are appropriate. The Congressional mandate, however, does not allow OSHA this flexibility. In addition, participants addressed this issue and the consensus was that the provisions of the standard were inextricably intertwined and they could not be considered separately without adversely affecting the contemplated effectiveness of the rule.

1.5 LIST OF REGULATED CHEMICALS AND THRESHOLD QUANTITIES

In establishing the list of substances to be regulated under the process safety management rule, OSHA carefully considered the potential for catastrophic events posed by a large number of chemicals. In order to select chemicals with catastrophic potential, OSHA consulted the lists developed by the Environmental Protection Agency and Department of Transportation and various states with regulatory experience in this area; namely Delaware and New Jersey. In developing the list of covered substances, OSHA also reviewed materials on this subject developed by the World Bank, the European Economic Community (the Seveso Directive), the National Fire Protection Association and ORC. While all chemicals on the list do not have equal catastrophic potential, OSHA addressed this issue in two ways. It developed appropriate thresholds for each of these chemicals by consulting with the sources above and relying on its own expertise, and it developed the flexible performance-oriented approach of the standard by mandating a process hazard analysis which will itself indicate the necessary safety precautions to take according to the incidence of use in a particular industrial setting.

1.6 EFFECTIVENESS RATE OF PSM STANDARD

OMB claimed that the effectiveness of OSHA's approach is uncertain. In its Preliminary Regulatory Impact Analysis (PRIA), OSHA claimed that after the standard had been in effect for five years, injuries and illnesses resulting from potentially catastrophic incidents would be reduced by at least 80 percent. This effectiveness rate is consistent with that used in other OSHA Regulatory Impact Analyses, such as Electrical Safety-Related Work Practices (Final Rule, 55 Fed. Reg. at 32011, August 6, 1990, Regulatory Impact Assessment); Control of Hazardous Energy Source (Lockout/Tagout) (85 percent, Final Rule, 54 Fed. Reg. at 36685, September 1, 1990, Regulatory Impact Analysis); Permit Required Confined Spaces; Notice of Proposed Rulemaking (54 Fed. Reg. at 24097, June 5, 1989, Benefits); and Hearing Conservation (Final Regulatory Analysis of the Hearing Conservation Amendment, U.S. Department of Labor, January 1981, Chapter III-27; benefits of 85 percent at equilibrium from the hearing conservation amendment).

Rulemaking participants also acknowledged that the process safety management standard will be substantially effective in improving safety.

1.6.1 Cost Impacts of the PSM Standard

OMB also indicated that the costs may be higher than estimated and may adversely affect profitability. The PRIA predicted that compliance with the proposed standard would cost $638 million in direct annualized gross costs (estimated per year for a ten year period). These estimates were based in large part on the Kearney/Centaur Report, "Proposed OSHA Rule for the Process Hazards Management of Highly Hazardous Chemicals: An Industry Profile, Cost Assessment and Benefits Analysis." A number of commenters believed that the PRIA had underestimated the costs of complying with the proposed standard. In response to comments in the record, OSHA updated and refined the Kearney/Centaur industry profile, its estimates of current industry compliance with the proposed process safety management standard, and the estimate of the number of processes per establishment that would be covered by the standard (Regulatory Impact Analysis, Chapter V). This resulted in the calculation of

14 OSHA and EPA PSM Requirements

increased costs of compliance with the process safety management standard.

The final Regulatory Impact Analysis (RIA) predicts gross costs of $863.5 million/year during the first five years that the standard is in effect (as opposed to the $638 million estimated by the PRIA; See RIA, Chapter IV-11) and $390.1 million/year during the next five years. In order to better understand the true costs associated with the process safety management rule, the gross cost of compliance must then be adjusted downward to account for the many benefits of the standard, such as increased productivity, decreased property damage, and decreased fatalities and injuries. When these offsets are taken into account, OSHA predicts that the standard will cost approximately $143.5 million per year for the first five years. Cost savings are expected to exceed direct costs for most industry groups in years six through ten.

OSHA also looked at the effect of the costs of compliance on the profitability of the affected industries and found that, assuming that the affected companies would not pass on any of the costs of compliance to customers (a worst case assumption), in the first five years compliance with the process safety management rule might decrease profits anywhere from .09 percent to 15.7 percent depending on the industry group. Worst-case profit impacts would average 1.1 percent for large establishments and 3.2 percent for small establishments. Therefore the final figures show that the standard is not only economically feasible, but it will not unreasonably affect the profitability of the affected industries and is well within the mandates of the CAAA, the Occupational Safety and Health Act and Executive Order 12291.

1.6.2 Impact on Small Businesses

OMB also believed that the process safety management rule might have high costs and few benefits for small business employers and, therefore, could be anticompetitive. The final RIA indicates the gross costs of complying with the process safety management standard will be $863.5 million/year during the first five years that the standard is in effect and $390.1 million/year during the next five years for all industry. Of this total gross cost of compliance, small business will bear approximately $88.9 million/year during the first five years that the standard is in effect and $33.0 million/year during the next five years. While accounting for approximately 10 percent of costs, small firms will realize considerable benefits from compliance; 21 percent of fatalities avoided and 9 percent of lost-workday injuries avoided will occur in small establishments.

OSHA's estimates for small-firm costs declined in the final impact analysis after incorporating the ideas of inventory reduction and a learning-curve effect during compliance. A small business might reduce the potential hazard by purposely controlling its on-site inventory of highly hazardous chemicals by ordering more frequent, smaller shipments so that it does not exceed the threshold for coverage set forth in the rule. Also, small businesses may segregate their inventory by dispersing the storage around the worksite so that the release of a highly hazardous chemical from one storage area would not cause the release of the other inventory stored on site. This remote storage approach would also be a feasible alternative. Moreover, small business employers who use batch processes may be able to use a generic approach to the required process hazards analysis which would help to further reduce the estimated cost of compliance. For example, a generic process hazard analysis of a representative batch might be used where there are only small changes in the process chemistry and this is documented for the range of batch processes.

Also, as a general rule, small business employers have greater flexibility within their workplaces than do large businesses. Employees may be trained to do more than one job and have a greater understanding of the interrelationship of the different factors that can adversely affect the process and produce a potentially catastrophic incident.

1.6.3 Sunset Provision Unnecessary

OMB also thought that OSHA should consider a sunset provision in the final rule that would cause the rule to expire after five years if it does not have the intended effect of providing significant reductions in the number of workplace accidents associated with hazardous chemicals. In its proposal, OSHA did not propose a specific timeframe for compliance. The final OSHA rule, because of feasibility considerations, does not become fully effective for five years and if the comment is read literally, the rule might "set" before it was fully implemented. Therefore, the five-year "sunset" timeframe would not be compatible with the

16 OSHA and EPA PSM Requirements

process safety management regulatory framework. The process safety management standard is based on the Occupational Safety and Health Act and the CAAA. The CAAA does not contemplate a "sunset" provision and this is probably because the chemicals which this standard regulates are intrinsically hazardous and the hazard will not go away as long as these chemicals are being used in industrial processes.

Even if the OMB comment were read to mean that OSHA should consider a sunset provision five years after the rule becomes effective, there is nothing to support the inclusion of a sunset provision in the final rule. The agency firmly believes that the final rule will be highly effective and will significantly reduce workplace accidents and injuries. Therefore, it would be arbitrary and contrary to the record evidence for the agency to include a sunset provision in the final rule. Moreover, it is questionable whether this approach (i.e., a sunset provision) is consistent with the procedural framework of the Occupational Safety and Health Act, which directs the Secretary to use specified procedures to amend or revoke a standard adopted under the Act. These procedures include public notice and an opportunity for the public to file comments and objections and to request a public hearing on the proposed amendments or revocation (29 U.S.C. 655).

It is possible that after the process safety management rule has been in effect for a while, however, facts may emerge to indicate that there is a need to change the regulation (e.g., safety prevention provisions, highly hazardous chemical lists, etc.). If such facts emerge, either based on safety experience under the rule or on an OSHA retrospective study of the costs and benefits of the rule, the Agency might then consider amending the regulation to make it more effective. This would be done under section 6 of the Occupational Safety and Health Act, perhaps with the assistance of other potentially relevant statutes such as the Alternative Dispute Resolution Act (Pub. L. 101-552) and the Negotiated Rulemaking Act (Pub. L. 101-648). Under any of these vehicles, however, interested persons would be given a chance to comment and present evidence on all relevant issues, a safeguard that might be missing if a sunset provision were used.

Chapter 2

Process Safety Management Regulatory Framework

2.1 OVERVIEW OF THE PSM STANDARD

For decades, the Occupational Health and Safety Administration (OSHA) has functioned as the worker safety police - setting safety standards, carrying out inspections, and whenever necessary, levying fines and penalties for noncompliance with OSHA regulations. A primary focus of OSHA regulation has been on companies that produce or use hazardous chemicals in the workplace. OSHA introduced its Hazard Communication Standard to require employers to inform employees about the presence of hazardous substances in the workplace. The HazCom program requires that each company obtain information about all hazardous chemicals on the premises, label all containers, develop a written hazard communication plan and inventory, explain to all employees the dangers of chemicals present at the company, and train employees to protect themselves from chemical exposure.

OSHA has taken the HazCom program a step further by introducing a new set of regulations aimed specifically at hazards associated with highly hazardous chemicals. In May 1992, OSHA's new "Process Safety Management Standard (PSM Standard)" went into effect. In February 1992, OSHA issued its final regulations regarding the process safety management (PSM) Standard. 57 Fed. Reg. 6403 (Feb. 24, 1992), codified at 29 CFR Section 1910.119.

The standard mainly applies to manufacturing industries, particularly those involved with chemicals, transportation equipment, and fabricated metal products. In each industry, the PSM Standard applies to those companies with processes involving any of more than 130 listed toxic and reactive chemicals above threshold quantities. For a listing of each toxic and reactive highly hazardous chemical and its threshold quantity refer to **Appendix 2**.

18 OSHA and EPA PSM Requirements

2.1.1 Impetus Behind the New PSM Standard

OSHA saw a pressing need for the new PSM Standard in light of the ongoing occurrence of serious accidents at chemical processing plants. Perhaps the single most vivid example of the dire consequences of improper management of hazardous chemicals is the Bhophal disaster in 1984. Several other catastrophes have taken place in the U.S. during the past decade, which emphasized the need for tighter regulatory control over hazards associated with processes that use or produce highly hazardous chemicals. The following is a list of some of the major accidents that have occurred in recent years:

June 1991: Tenneco, Inc.'s phosphorus chemical plant in Charleston, SC explodes, killing 6;
May 1991: IMC Fertilizer-Angus Chemical Co.'s nitropariffins plant in Sterlington, LA explodes, killing 8 and injuring 128;
March 1991: Union Carbide's petrochemicals plant in Seadrift, TX explodes, killing 1;
July 1990: BASF Corp.'s chemical plant in Cincinnati, OH explodes, killing 2;
July 1990: Arco Chemical Co.'s petrochemical plant in Channelview, TX explodes, killing 17;
October 1989: Phillips Petroleum Co.'s plastics plant in Houston, TX explodes, killing 23 and injuring some 350.

2.1.2 Performance-Oriented Approach

Investigations of major accidents concluded that most of the incidents were caused by one or more of four key factors: (1) technology; (2) management systems; (3) human error; and (4) external events. Human error and inadequate management systems were the most significant contributing factors to major plant accidents. Recognizing that human error can be minimized, and better management systems can be implemented to prevent such disasters, OSHA's new PSM Standard sets forth a performance-based approach to managing hazards related to use and production of highly hazardous chemicals. The emphasis on performance-based management controls largely developed as a result of OSHA's 1987 report on Chem-SEP, a special inspection program of 40 facilities that was carried out following the Bhophal disaster.

Chapter 2 - PSM Regulatory Framework   19

The performance-oriented approach of the PSM Standard differs from previous worker safety regulations that prescribe use of various types of equipment or specify certain emergency response actions. The PSM Standard establishes goals and describes basic elements of a program designed to fulfill these goals, however, gives companies flexibility in choosing the methods that will best serve the company's particular situation and needs.

The precautionary measures implemented under the new regulations are expected to reduce the risk of catastrophic fires and explosions by 80percent, saving 264 lives and preventing some 1,500 injuries per year. The new standard is expected to cost industry about $899 million per year for the first five years of the program, and another $406 million a year thereafter. OSHA, however, claims that these costs will be offset by the savings realized in accident prevention and productivity increases, which the agency has estimated at $720 million and $1.44 billion over the same periods. (*See* 55 Fed. Reg. 29,150 (July 17, 1990) (proposed rule); 57 Fed. Reg. 6403 (Feb. 24, 1992) (final rule).)

2.2   BASIC PSM STANDARD REQUIREMENTS

The term "process" in the regulations is broadly defined to include any activity involving a highly hazardous chemical, including using, storing, manufacturing, handling or moving such chemicals at the site, or any combination of these activities. Under the new PSM Standard, employers must complete a process hazard analysis (PHA) for all processes involving any listed toxic and reactive chemical. The PHA is a review of what could go wrong and what steps must be taken to safeguard against releases of any highly hazardous chemicals. Employers must identify those processes that pose the greatest risks and complete a process hazard analaysis for those processes first. Under the PSM Standard, employers must complete all PHAs as soon as possible, and at least 25 percent of all processes must be analyzed by May 26, 1994. By May 26 of each subsequent year, an additional 25percent of all processes must be evaluated so that by May 26, 1997, all PHAs will be completed. The Process Hazard Analysis is easily the most time-consuming aspect of complying with the new PSM Standard. Companies covered by the new standard, are presently struggling to meet the compliance deadline for the first PHA phase (May 1994).

20 OSHA and EPA PSM Requirements

In addition to the completion of PHAs for all processes, the PSM Standard mandates written operating procedures, employee training, pre-startup safety reviews, evaluation of mechanical integrity of critical equipment, and written procedures for managing change. The PSM standard specifies a permit system for "hot work" (work involving electric or gas welding, cutting, brazing, or similar flame or spark-producing operations), investigations of incidents involving releases or near-misses of covered chemicals, emergency action plans, and compliance audits every three years.

The final regulations for the PSM Standard were issued in May 1992, and are described in detail, with guidance for compliance, throughout this book. The final process safety management standard contains the following paragraphs:

Purpose
Application: Paragraph (a)
Definitions: Paragraph (b)
Employee participation: Paragraph (c)
Process safety information: Paragraph (d)
Process hazards analysis: Paragraph (e)
Operating procedures: Paragraph (f)
Training: Paragraph (g)
Contractors: Paragraph (h)
Pre-startup safety review: Paragraph (i)
Mechanical integrity: Paragraph (j)
Hot work permit: Paragraph (k)
Management of change: Paragraph (l)
Incident investigation: Paragraph (m)
Emergency planning and response: Paragraph (n)
Compliance safety audit: Paragraph (o)
Trade secrets: Paragraph (p)

2.3 APPLICATION OF THE PSM STANDARD: PARAGRAPH (a)

The application section in paragraph (a) of the Rule specifies those types of highly hazardous chemicals covered by the regulations. The application section also includes processes involving certain specified highly hazardous chemicals at or above a stated threshold which is listed in appendix A to the Rule and processes involving flammable liquids or gases on site in one location in quantities of 10,000 pounds or greater (with two exceptions discussed below). Additionally, OSHA excludes retail facilities, oil and gas well drilling and servicing operations and normally unmanned remote facilities from the standard.

2.3.1 Meaning of the Term "Processes"

Before considering the application provisions in detail, its is important to understand OSHA's use of the plural word "processes" in the application paragraph of the regulations. Rulemaking participants questioned whether the use of the word "processes" meant that the amount of highly hazardous chemical used at a plant must be aggregated to meet the threshold for coverage even though the amount of highly hazardous chemicals used at any one location might be less than the threshold amount or the amount of highly hazardous chemical in use might be divided among remote processes.

Rulemaking participants also wondered whether the proposal required that the highly hazardous chemical threshold quantity be aggregated over a period of time or whether it must be present at one point in time to be covered by the proposal.

OSHA addressed this concern in its November 1, 1990, Federal Register notice in Issue 2 (55 Fed. Reg. 46075). OSHA's view at that time was that if a plant exceeded the threshold quantity of a listed chemical but the chemical was used in smaller quantities around the plant and was not concentrated in one process or in one area, then the chance of a catastrophic release of the threshold quantity would be remote due to the reduced availability of a concentrated amount of the chemical in one location. However, OSHA requested comment on the point at which a chemical should be considered in its aggregate due to the proximity of the sites at which it was being used in a plant.

While a few participants indicated that the amounts of a highly hazardous chemical used at various sites around the plant should all be counted toward the threshold amount for coverage, most participants who discussed this issue noted that the threshold quantity should not be aggregated. They agreed that highly hazardous chemicals in less than threshold quantities distributed in several processes would not present as great a risk of catastrophe as the threshold quantity in a single process. OSHA felt that the potential hazard of a catastrophic release exists when the highly hazardous chemical is concentrated in a process and, therefore,

22 OSHA and EPA PSM Requirements

agreed with these commenters. OSHA clarified the language contained in the application paragraph to reflect its intent that coverage is triggered by a specified threshold quantity of an Appendix A substance being used in a single process. This language also clarifies the fact that the presence of a threshold quantity of a highly hazardous chemical in a process is to be at one point in time; not aggregated over a period of time.

2.3.2 List of Highly Hazardous Chemicals

In the application section, a process would be covered if it involved a toxic or reactive highly hazardous chemical listed in Appendix A, at or above a specified threshold quantity. Appendix A is a compilation of highly hazardous chemicals that could cause a serious chemical accident, by toxicity or reactivity, and a consequent potential danger to employees in a workplace. The Appendix A list has been drawn from a variety of relevant sources which include: The New Jersey "Toxic Catastrophe Prevention Act," the State of Delaware's "Extremely Hazardous Substances Risk Management Act," the World Bank's "Manual of Industrial Hazard Assessment Techniques," the Environmental Protection Agency's "Extremely Hazardous Substance List," the Seveso Directive on major accident hazards of certain industrial activities (82/501/EEC), the United Kingdom's "A Guide to the Control of Industrial Major Accident Hazards Regulations 1984," the American Petroleum Institute's RP 750, "Management of Process Hazards," the National Fire Protection Association's NFPA 49, "Hazardous Chemicals Data," and the Organization Resources Counselors, Inc.'s "Recommendations for Process Hazards Management of Substances with Catastrophic Potential."

Every chemical listed in Appendix A is on at least one list compiled by these agencies and organizations as warranting a high degree of management control due to its extremely hazardous nature. Most of the chemicals are on several lists. Not every list contains the same chemicals or quantities. Based on a review of these sources, OSHA has sought to include those toxic and reactive chemicals it believes are most significant in potentially becoming a catastrophic event. OSHA has also sought to develop a reasonable listing of threshold quantities which, when used in a process, would invoke coverage of the standard. Those Appendix A highly hazardous chemicals which are highly reactive or explosive-type chemicals have been drawn from chemicals listed in the National

Fire Protection Association (NFPA) document, NFPA 49, "Hazardous Chemicals Data" and cross-referenced with other sources mentioned above. The agency decided to include substances with the two highest or most dangerous reactivity ratings from NFPA 49 because these chemicals present the most severe exposure potential to workers. These substances, which are rated 3 or 4 by NFPA 49, are those which are capable of undergoing detonation or explosive decomposition. These are the substances that can generate the most severe blast or shock wave, and can cause fragmentation of piping, vessels and containers, as well as cause serious damage to buildings and structures.

2.3.3 Minimum Threshold Levels

The minimum threshold quantities for the highly reactive chemicals covered by the standard have been determined by calculating the amount of material needed to propagate a blast wave that creates an overpressure of 2.3 p.s.i. (15.85 kPa) to a flat surface perpendicular to the direction of the blast wave at a distance of 100 meters from the point of origin. This approach is similar to that used by the State of Delaware. In determining threshold quantities for toxic chemicals, OSHA used the Turner-described Gaussian dispersion model. This approach is similar to that used by the State of Delaware. Both OSHA and Delaware made the following assumptions: Average conditions of 4.3 m/sec. wind speed and D stability with urban dispersion coefficients; continuous steady-state release for one hour; no liquid pools; all released chemicals in vapor or gaseous state; chemical release is at ambient temperature and at ground level; chemical gas or vapor cloud is neutrally buoyant; and no design features prevent downwind dispersion. The calculated threshold quantities were rounded off by OSHA to further simplify the standard.

The lowest threshold quantity that the agency has used is 100 pounds (45.4 kg) for the most hazardous of the chemicals listed. The OSHA threshold quantities are the same or somewhat greater than the Delaware "sufficient quantity level" (threshold quantity) due to the rounding up by the Agency for the vast majority of the toxic chemicals listed. This has been done to simplify the application of this PSM rule and also in recognition that the agency has other standards which address the hazards of lower quantities of toxic materials in the workplace.

24 OSHA and EPA PSM Requirements

2.3.4 Role of EPCRA and CAAA Chemical Lists

Under section 302 of the Superfund Amendments and Reauthorization Act (SARA) also known as the Emergency Planning and Community Right-to-Know Act (EPCRA, 42 U.S.C. 11001 *et seq.*), EPA was required to publish a list of extremely hazardous substances (EHSs) with threshold planning quantities which would trigger planning in states and local communities (52 Fed. Reg. 13378). EPA's EHS list is quite extensive (more than 300 hazardous substances) and serves as an emergency response planning list directed toward addressing hazards to the public and the environment.

Section 304 of the Clean Air Act Amendments (CAAA), paragraph (b), List of Highly Hazardous Chemicals, mandates that: The Secretary (of Labor) shall include as part of such standard (Chemical Process Safety Standard) a list of highly hazardous chemicals, which include toxic, flammable, highly reactive and explosive substances. The paragraph further indicates that the Secretary may include those chemicals listed by the Environmental Protection Agency under section 302 of the Emergency Planning and Community Right-to-Know Act of 1986. Further the CAAA did not anticipate that even EPA would adopt the whole EHS list for the purpose of prevention of accidental chemical releases. Section 301(r) indicated that EPA's first list must contain no less than 100 substances which may be from the EHS list. EPA's 301(r) list is not a planning tool but rather a list that requires covered plants to develop comprehensive Risk Management Plans.

While OSHA considered this list, it does not consider all of the substances on the EHS list to present a potential catastrophic situation for employees in workplaces within its jurisdiction. Therefore, OSHA believes it has acted reasonably and appropriately in evaluating a variety of chemical lists including the EHS list in order to identify those highly hazardous chemicals which present a potential catastrophic threat to employees. These events typically include toxic releases, fires and explosions as opposed to potential environmental threats such as spillage of a pesticide.

Several participants in the rulemaking advised OSHA that certain chemicals which appeared in Appendix A, including dimethyl sulfide, isopropyl formate, and methyl disulfide had been deleted from EPA's EHS list based on a reconsideration of the data and a determination that the data did not support the

Chapter 2 - PSM Regulatory Framework 25

inclusion of the chemicals on the EHS list. OSHA also deleted these chemicals from its list since a redetermination had been made that data and information available did not support their inclusion on the EPA list.

A number of rulemaking participants felt OSHA should provide a technical basis for the Appendix A list and its threshold quantities. Some participants noted that if no published technical basis existed then it would be difficult to add chemicals at a later time. OSHA's view is that a means for adding highly hazardous chemicals to Appendix A in the future can be considered at such time as the need arises.

2.3.5 Exemptions for Hydrocarbon Fuels

Various concerns were expressed during the rulemaking over the wording of the proposed exemption for hydrocarbon fuels that are present in quantities greater than 10,000 pounds, but are not part of a process. Examples of these would be propane or oil used for comfort heating and gasoline or diesel fuel for use in industrial vehicles. To remedy this confusion, the Organization Resources Counselors (ORC) recommended that the language of the exemption be amended to read: "Hydrocarbon fuels used solely as a fuel at a facility which is not otherwise covered by this rule." OSHA agreed and changed the final provision to clarify its intent not to exclude from coverage hydrocarbon fuels used for process related applications such as furnaces, heat exchangers and the like at facilities covered by this rule. This exemption is contained in paragraph (a)(1)(ii)(A) and exempts from coverage:
"Hydrocarbon fuels used solely for workplace consumption as a fuel (e.g., propane used for comfort heating or gasoline used for vehicle fueling), if such fuels are not a part of a process containing another highly hazardous chemical covered by this standard."

2.3.6 Flammable Liquids Exemption

The second exemption concerns flammable liquids stored or transferred which are kept below their atmospheric boiling point without benefit of chilling or refrigeration (paragraph (b)(1)(ii)(B)). OSHA did not believe that the flammable liquids as described in the exemption have the same potential for a catastrophe. An OSHA standard already regulates the treatment of

26 OSHA and EPA PSM Requirements

the exempted flammable liquids (29 CFR 1910.106, flammable and combustible liquids).

While many rulemaking participants supported the exemption concerning flammable liquids stored or transferred which are kept below their atmospheric boiling point without benefit of chilling or refrigeration, they recommended that OSHA clarify the exemption by using established language from its standard concerning flammable and combustible liquids. For example, the American Petroleum Institute concluded: "OSHA's phrase 'atmospheric boiling point' introduces unnecessary problems in applying this important exemption to various complex substances such as crude oil which do not have precise boiling points. OSHA has previously resolved this problem by providing definitions for 'atmospheric tank' and 'boiling point' in subpart H--1910.106(a)(2) and (a)(5)."

OSHA agreed with this suggestion concerning the use of existing definitions. Therefore, the wording of paragraph (a)(1)(ii)(B) was clarified by adding existing language from OSHA's standard for flammable and combustible liquids, 29 CFR 1910.106, "atmospheric tank" and "boiling point," and providing a definition for these terms in lieu of the term "atmospheric boiling point."

2.3.7 Other Exemptions

Certain other exemptions are contained in the application paragraph of the process safety management rule. These exemptions included: retail facilities; oil and gas well drilling and servicing; and normally unmanned remote facilities. With respect to the exclusion of retail facilities and normally unmanned remote facilities, OSHA believed that such facilities did not present the same degree of hazard to employees as other workplaces covered by the proposal. Therefore OSHA should not require a comprehensive process safety management system in addition to other applicable OSHA standards addressing flammable and combustible liquids, compressed gases, hazard communication, etc., for retail facilities and unmanned remote facilities.

Certainly, highly hazardous chemicals may be present in both types of work operations. However, OSHA believed that chemicals in retail facilities are in small volume packages, containers and allotments, making a large release unlikely. In normally unmanned remote facilities (called "normally unoccupied remote facilities" in final paragraph (b)), the likelihood of an uncontrolled release injuring or killing employees is effectively reduced by isolating the process from employees. OSHA believes that the present OSHA standards contained in subpart H, such as 29 CFR 1910.101, compressed gases, and 29 CFR 1910.106 flammable and combustible liquids and in part 1910, subpart Z, toxic and hazardous substances, adequately address the chemical hazards presented in these work operations. OSHA excluded normally unmanned remote facilities because the agency believed that such facilities pose a reduced likelihood of releases that could harm employees.

OSHA also excluded oil and gas well drilling and servicing operations because OSHA had already undertaken rulemaking with regard to these activities (48 Fed. Reg. 57202). OSHA felt that oil and gas well drilling and servicing operations should be covered in a standard designed to address the uniqueness of that industry.

Finally, various rulemaking participants requested special consideration for their processes or exemption from the standard. For example, concern was expressed by participants who conduct batch processing operations regarding their ability to comply with the standard due to the dynamic nature of batch processing. With respect to this, the Synthetic and Organic Chemical Manufacturers Association (SOCMA) stated: "Batch processes are distinct from continuous operations in that a continuous operation has a constant raw material feed to the process and continual product withdrawal from the process. A batch process has an intermittent introduction of frequently changing raw materials into the process, varying process conditions imposed on the process within the same vessel depending on the product being manufactured. Consequently under the process safety rule as proposed, a batch processor will be required to perform a process hazard analysis each time an order comes in for a chemical that may differ only slightly from the one previously produced. A batch processing plant is in a constant state of change, always being adapted for different projects. It is not unusual for a batch processor to have a different plant configuration weekly." SOCMA suggested that batch processors be given the flexibility to do a process hazard analysis that is representative of many similar batches.

However, other rulemaking participants who have been involved in running both continuous processing and batch processing indicated that the standard for process safety management is appropriate for batch processing. The Chemical

28 OSHA and EPA PSM Requirements

Manufacturers Association (CMA) stated: "CMA does not believe that facility owners/operators with batch processes should be exempted from complying with the proposed PSM standard .... The key question is whether the hazardous material is present in an amount at or above the threshold quantity. If the answer to this question is yes, then the provisions of the proposed standard should apply. CMA companies have extensive experience handling listed materials both in batch and continuous facilities. CMA supports applying the provisions of the proposed standard to any facility (batch or continuous) where the threshold quantities are exceeded."

OSHA agreed that the key question for coverage was whether the highly hazardous chemical is present in an amount at or above the threshold. However, OSHA acknowledged the concern of SOCMA regarding the potential difficulty of conducting a separate process hazard analysis for each variation of a batch process. OSHA, therefore, accepted SOCMA's suggestion concerning the development of a generic process hazard analysis which is representative of similar batches.

2.4 DEFINITIONS: PARAGRAPH (b)

Paragraph (b) of the regulations contains the definitions of terms used for the PSM Standard. The regulations contains definitions for the following terms: Atmospheric tank, boiling point, catastrophic release, facility, highly hazardous chemical, hot work, normally unoccupied remote facility, process, replacement in kind, and trade secret. These definitions are reproduced below:

(b) Definitions.

Atmospheric tank means a storage tank which has been designed to operate at pressures from atmospheric through 0.5 p.s.i.g. (pounds per square inch gauge, 3.45 kPa).

Boiling point means the boiling point of a liquid at a pressure of 14.7 p.s.i.a. (pounds per square inch absolute). For the purposes of this section, where an accurate boiling point is unavailable for the material in question, or for mixtures which do not have a constant boiling point, the 10 percent point of a distillation performed in accordance with the Standard Method of Test for Distillation of

Petroleum Products, ASTM D-86-62, may be used as the boiling point of the liquid.

Catastrophic release means a major uncontrolled emission, fire, or explosion, involving one or more highly hazardous chemicals, that presents serious danger to employees in the workplace.

Facility means the buildings, containers or equipment which contain a process.

Highly hazardous chemical means a substance possessing toxic, reactive, flammable, or explosive properties and specified by paragraph (a)(1) of this section.

Hot work means work involving electric or gas welding, cutting, brazing, or similar flame or spark-producing operations.

Normally unoccupied remote facility means a facility which is operated, maintained or serviced by employees who visit the facility only periodically to check its operation and to perform necessary operating or maintenance tasks. No employees are permanently stationed at the facility. Facilities meeting this definition are not contiguous with, and must be geographically remote from all other buildings, processes or persons.

Process means any activity involving a highly hazardous chemical including any use, storage, manufacturing, handling, or the on-site movement of such chemicals, or combination of these activities. For purposes of this definition, any group of vessels which are interconnected and separate vessels which are located such that a highly hazardous chemical could be involved in a potential release shall be considered a single process.

Replacement in kind means a replacement which satisfies the design specification.

Trade secret means any confidential formula, pattern, process, device, information or compilation of information that is used in an employer's business, and that gives the employer an opportunity to obtain an advantage over competitors who do not know or use it. Appendix D contained in 29 CFR 1910.1200 sets out the criteria to be used in evaluating trade secrets.

2.5 HOT WORK PERMIT: PARAGRAPH (k)

In paragraph (k)(1), OSHA requires the employer to issue a permit for all hot work operations. The purpose of this provision is to assure that the employer is aware of the hot work being performed, and that appropriate safety precautions have been taken prior to beginning the work. The permit reminds the person performing the work of the steps necessary to perform the work safely; and if the hot work is performed on or near a covered process, then a permit should be required regardless of who is present. Paragraph (k)(1) of the PSM rule reads as follows: "The employer shall issue a hot work permit for hot work operations conducted on or near a covered process."

Paragraph (k)(2) requires the permit to certify that the fire prevention and protection requirements contained in 29 CFR 1910.252(a) have been implemented prior to beginning the hot work operations; indicate the date authorized for the hot work; and identify the equipment or facility on which the hot work is to be performed. It also requires the permit to be kept on file until completion of the hot work.

2.5.1 Hot Work Permit Compliance Checklist

1910.119 (k): HOT WORK PERMIT

I. PROGRAM SUMMARY

The intent of this paragraph is to require employers to control, in a consistent manner, nonroutine work conducted in process areas. Specifically, this subparagraph is concerned with the permitting of hot work operations associated with welding and cutting in process areas.

Minimum requirements include: that the employer issue a hot work permit for hot work operations conducted on or near a covered process and that hot work permits shall document compliance with the fire prevention and protection requirements of 29 CFR 1910.252(a).

II. QUALITY CRITERIA REFERENCES

    A.   1910.119(k):   Hot Work Permit
    B.   1910.252(a):   Fire Prevention and Protection

III. VERIFICATION OF PROGRAM ELEMENTS

	Criteria Reference	Met Y/N
A. Records Review	.119(k)(1)	
1. Have hot work permits been issued for all hot work operations conducted on or near a process covered by this standard? FIELD NOTE REFERENCE(S):	.119(k)(1)	
2. Do the hot work permits indicate the date(s) authorized for hot work performed? FIELD NOTE REFERENCE(S):	.119(k)(2)	
3. Do the hot work permits describe the object on which the hot work is to be performed? FIELD NOTE REFERENCE(S):	.119(k)(2)	

32 OSHA and EPA PSM Requirements

4. Have the hot work permits been kept on file until the hot work operations were complete? FIELD NOTE REFERENCE(S):	.119(k)(2)
5. Have the hot work permits identified openings, cracks and holes where sparks may drop to combustible materials below? FIELD NOTE REFERENCE(S):	.252(a)(2)(i)
6. Have the hot work permits described the fire extinguishers required to handle any emergencies? FIELD NOTE REFERENCE(S):	.252(a)(2)(ii)
7. Have the hot work permits assigned fire watchers whenever welding is performed in locations where other than a minor fire might develop? FIELD NOTE REFERENCE(S):	.252(a)(2)(iii)
8. Are the hot work permits being authorized, preferably in writing, by the "individual" responsible for all welding and cutting operations? Is authorization preceded by site inspection and designation of appropriate precautions? FIELD NOTE REFERENCE(S):	.252(a)(2)(iv) & .252(a)(2)(xiii)(A)
9. Have the hot work permits described precautions associated with combustible materials on floors or floors, walls, partitions, ceilings or roofs of combustible construction? FIELD NOTE REFERENCE(S):	.252(a)(2)(v) & .252(a)(2)(ix)
10. Has hot work permitting been successful in prohibiting welding in unauthorized areas, in sprinklered buildings while such protection is impaired, in the presence of explosive atmospheres, and in storage areas for large quantities of readily ignitable materials? FIELD NOTE REFERENCE(S):	.252(a)(2)(vi)

Chapter 2 - PSM Regulatory Framework 33

11. Have the hot work permits required relocation of combustibles where practicable and covering with flameproofed covers where not practicable? FIELD NOTE REFERENCE(S):	.252(a) (2) (vii)	
12. Have hot work permits identified for shutdown any ducts or conveyors systems that may convey sparks to distant combustibles? FIELD NOTE REFERENCE(S):	.252(a) (2) (viii)	
13. Have hot work permits required precautions whenever welding on components (e.g., steel members, pipes, etc,) that could transmit heat by radiation or conduction to unobserved combustibles? FIELD NOTE REFERENCE(S):	.252(a) (2)(x) & .252(a) (2) (xii)	
14. Have hot work permits identified hazards associated with welding on walls, partitions, ceilings or roofs with combustible coverings or welding on walls or panels of sandwich-type construction? FIELD NOTE REFERENCE(S):	.252(a) (2)(xi)	
15. Has management established areas and procedures for safe welding and cutting based on fire potential? FIELD NOTE REFERENCE(S):	.252(a) (2) (xiii)	
16. Has management designated the "individual" responsible for authorizing cutting and welding operations in process areas? FIELD NOTE REFERENCE(S):	.252(a) (2) (xiii) (B)	
17. Has management ensured that welders, cutters and supervisors are trained in the safe operation of their equipment? FIELD NOTE REFERENCE(S):	.252(a) (2) (xiii) (C)	

34 OSHA and EPA PSM Requirements

18. Has management advised outside contractors working on their site about all hot work permitting programs? FIELD NOTE REFERENCE(S):	.252(a) (2) (xiii) (D)
19. Has the Supervisor determined if combustibles are being protected from ignition prior to welding by moving them, shielding them, or scheduling welding around their production? FIELD NOTE REFERENCE(S):	.252(a) (2) (xiv) (A)(B) & (C)
20. Has the Supervisor, prior to welding, secured authorization from the responsible "individual" designated by management? FIELD NOTE REFERENCE(S):	.252(a) (2) (xiv) (D)
B. On-Site Conditions 1. Conduct checks of <u>current</u> welding and cutting operations to ensure compliance with the requirements of 1910.119(k) and 1910.252(a). The twenty items listed above in "Records Review" may serve as an audit checklist. A <u>management representative</u>, the "<u>individual</u>" <u>responsible</u> for welding operations and the <u>supervisor</u> should all be invited to participate in this on-site check. FIELD NOTE REFERENCE(S):	.119(k) & .252(a)
C. Interviews - Employees and Contractors 1. Based on interviews with a representative number of maintenance and contractor employees, has the Supervisor visited welding work operations to verify that: • Welders have approval for safe go ahead prior to welding? • Fire protection and extinguishing equipment is properly located at the work site? • Fire watches are functional, where required? FIELD NOTE REFERENCE(S):	.252(a) (2) (xiv) (E), (F) & (G)

Chapter 2 - PSM Regulatory Framework 35

2. Based on interviews with a representative number of maintenance and contractor employees, have hot work permits been issued for all hot work operations conducted on or near a process covered by this standard? FIELD NOTE REFERENCE(S):	.119(k) (1)
3. Based on interviews with a representative number of maintenance and contractor employees, have the hot work permits been kept on file until the hot work operations were complete? FIELD NOTE REFERENCE(S):	.119(k) (2)
4. Based on interviews with a representative number of maintenance and contractor employees, have the hot work permits identified openings, cracks and holes where sparks may drop to combustible materials below? FIELD NOTE REFERENCE(S):	.252(a) (2)(i)
5. Based on interviews with a representative number of maintenance and contractor employees, have the hot work permits assigned fire watchers whenever welding is performed in locations where other than a minor fire might develop? FIELD NOTE REFERENCE(S):	.252(a) (2) (iii)
6. Based on interviews with a representative number of maintenance and contractor employees, are the hot work permits being authorized, preferably in writing, by the "individual" responsible for all welding and cutting operations? Is authorization preceded by site inspection and designation of appropriate precautions? FIELD NOTE REFERENCE(S):	.252(a) (2)(iv) & .252(a) (2) (xiii) (A)
7. Based on interviews with a representative number of maintenance and contractor employees, have the hot work permits described precautions associated with combustible materials on floors or floors, walls, partitions, ceilings or roofs of combustible construction? FIELD NOTE REFERENCE(S):	.252(a) (2)(v) & .252(a) (2)(ix)

36 OSHA and EPA PSM Requirements

8. Based on interviews with a representative number of maintenance and contractor employees, has hot work permitting been successful in prohibiting welding in: • Unauthorized areas? • Sprinklered buildings while such protection is impaired? • The presence of explosive atmospheres? • Storage areas for large quantities of readily ignitable materials? FIELD NOTE REFERENCE(S):	.252(a)(2)(vi)
9. Based on interviews with a representative number of maintenance and contractor employees, have the hot work permits required relocation of combustibles where practicable and covering with flameproofed covers where not practicable? FIELD NOTE REFERENCE(S):	.252(a)(2)(vii)
10. Based on interviews with a representative number of maintenance and contractor employees, have hot work permits identified for shutdown any ducts or conveyors systems that may convey sparks to distant combustibles? FIELD NOTE REFERENCE(S):	.252(a)(2)(viii)
11. Based on interviews with a representative number of maintenance and contractor employees, have hot work permits required precautions whenever welding on components (e.g., steel members, pipes, etc.) that could transmit heat by radiation or conduction to unobserved combustibles? FIELD NOTE REFERENCE(S):	.252(a)(2)(x) & .252(a)(2)(xii)
12. Based on interviews with a representative number of maintenance and contractor employees, have hot work permits identified hazards associated with welding on walls, partitions, ceilings or roofs with combustible coverings or welding on walls or panels of sandwich-type construction? FIELD NOTE REFERENCE(S):	.252(a)(2)(xi)

13. Based on interviews with a representative number of maintenance and contractor employees, has management established areas and procedures for safe welding and cutting based on fire potential? FIELD NOTE REFERENCE(S):	.252(a) (2) (xiii)
14. Based on interviews with a representative number of maintenance and contractor employees, has management designated the "individual" responsible for authorizing cutting and welding operations in process areas? FIELD NOTE REFERENCE(S):	.252(a) (2) (xiii) (B)
15. Based on interviews with a representative number of maintenance and contractor employees, has management ensured that welders, cutters and supervisors are trained in the safe operation of their equipment? FIELD NOTE REFERENCE(S):	.252(a) (2) (xiii) (C)
16. Based on interviews with contractors and contractor employees, has management advised outside contractors working on the site about all hot work permitting programs? FIELD NOTE REFERENCE(S):	.252(a) (2) (xiii) (D)
17. Based on interviews with a representative number of maintenance and contractor employees, has the Supervisor determined if combustibles are being protected from ignition prior to welding by moving them, shielding them, or scheduling welding around their production. FIELD NOTE REFERENCE(S):	.252(a) (2) (xiv) (A)(B) & (C)

2.6 TRADE SECRETS: PARAGRAPH (p)

Participants in the initial rulemaking expressed some concern that the proposed OSHA regulations did not appear to provide any trade secret protection. Employers are, however, protected under the U.S. Code, the Occupational Safety and Health Act and regulations promulgated under the Act. Federal law makes it a criminal offense for federal employees to disclose trade secret information that is not authorized by law (18 U.S.C. 1905). Section 15 of the Occupational Safety and Health Act (the Act) requires that all information reported to or obtained by a Compliance Safety and Health Officer (CSHO), in connection with any inspection or other activity, which contains or which might reveal a trade secret be kept confidential. Such information shall not be disclosed except to other OSHA officials concerned with the enforcement of the Act or, when relevant, in any proceeding under the Act. Other OSHA regulations further ensure the protection of trade secrets (29 CFR 1903.7(b) and 1903.9). And the OSHA Field Operations Manual further emphasizes this point by stating "it is essential to the effective enforcement of the Act that the CSHO and all OSHA personnel preserve the confidentiality of all information and investigations which might reveal a trade secret" (III-58). Moreover, trade secret information is specifically excluded from disclosure under the Freedom of Information Act (5 U.S.C. 552(b)(4)).

As a general matter, OSHA felt that there are relatively few bona fide trade secrets among the information that is required to be gathered under the PSM Standard. However, the addition of provisions to protect trade secrets were added to give employers with legitimate trade secret concerns adequate protection, while requiring that they withhold information only on the basis of sound, legal justification. Some rulemaking participants suggested that OSHA adopt the definition of "trade secret" used in the Hazard Communication Standard. Others suggested a more expansive or more limited definition.

OSHA has reviewed the definition of "trade secret" that is used in the Hazard Communication Standard (29 CFR 1910.1200) and decided to incorporate that definition of trade secret into the final PSM Standard. The agency felt that this definition of trade secret is broad enough to offer adequate protection to employers with legitimate trade secrets, is consistent with that used in the Restatement of Torts, and has the additional advantage of being uniform with that used in the Hazard Communication Standard so that many employers are already familiar with it. The final rule also incorporated Appendix D of the Hazard Communication Standard, which contains criteria to be used in determining whether material meets the definition of trade secret.

To further clarify its intent, OSHA also has specifically requires that the employer must make all relevant information available to those individuals involved in carrying out various information using and compiling activities required by the rule regardless of whether the information in question is considered a trade secret or not. OSHA felt that this was vital to the effective operation of the process safety management rule since it would be questionable how useful a compliance safety audit or a process hazard analysis could be if some of the information necessary to their completion were denied or delayed.

Thus, the language of this provision is written to emphasize the right to access this information, while permitting the employer to take reasonable steps, such as those described in the Hazard Communication Standard, to protect against the unauthorized disclosure of trade secrets to unauthorized third persons. Such steps include the signing of a confidentiality agreement. OSHA felt that employees and their representatives also may have the need to access such information. The final rule assures employees access to the process hazard analysis and other information required to be developed under the standard. Under certain circumstances, however, it might be appropriate to substitute more general information or to require some sort of a balancing of the need to know the information with the need to protect the employer. Therefore, the Agency is incorporating into the final rule the access procedures that were developed under the Hazard Communication Standard with the exception of 29 CFR 1910.1200(i)(13).

Section 1910.1200(i)(13) provides that "(n)othing in this paragraph shall be construed as requiring the disclosure under any circumstances of process or percentage of mixture information which is a trade secret." That section is not being incorporated into the process safety management trade secret provisions in recognition of the fact that employees are entitled to certain process information under the process safety management standard and this proces information may at times contain trade secret information. The Hazard Communication information access provisions should work well for information contained in the

40 OSHA and EPA PSM Requirements

process hazard analysis and other documents that contain trade secrets. Employers bear the burden of demonstrating that their trade secret claim is bona fide. The Agency will evaluate the appropriateness of that substantiation in the event that an employer denies a legitimate request for disclosure of the trade secret and a complaint is subsequently made to OSHA.

2.6.1 Trade Secrets Compliance Checklist

1910.119 (p): TRADE SECRETS

I. PROGRAM SUMMARY

The intent of this paragraph is to require employers to provide all information necessary to comply with the standard to personnel developing paragraphs (d), (e), (f), (n) and (o) without regard to possible trade secrets. In addition, employees and their designated representatives shall have access to trade secret information contained within documents required to be developed by the standard.

II. QUALITY CRITERIA REFERENCES

    A.   1910.119(p):   Trade Secrets
    B.   1910.1200:   Hazard Communication

III. VERIFICATION OF PROGRAM ELEMENTS	Criteria Reference	Met Y/N
A. Records Review		
1. Has all information necessary been provided to those persons responsible for compiling the process safety information (paragraph d), those assisting in development of the PHA (paragraph e), those responsible for developing the operating procedures (paragraph f), and those involved in incident investigations (paragraph m) and emergency planning and response (paragraph n), and compliance audits (paragraph o) been without regard to possible trade secret status of such information? FIELD NOTE REFERENCE(S):	.119(p)(1)	
2. Do employees and their designated representatives have access to trade secret information contained in the PHA and to other documents required to be developed by the standard, subject to the provisions set forth in 1910.1200(i)(1) through (i)(12)? FIELD NOTE REFERENCE(S):	.119(p)(3)	
B. On-site Conditions Not applicable.		

42   OSHA and EPA PSM Requirements

C. Interviews **Employees involved in specific duties:** 1. Based on interviews with a representative number of employees involved in compiling the process safety information, developing PHA's, developing operating procedures, investigating incidents, planning and responding to emergencies, and auditing compliance, has all information necessary been provided to them without regard to possible trade secret status of such information? FIELD NOTE REFERENCE(S):	.119(p)(1)	
Employees and Representatives: 2. Based on interviews with a representative number of employees and their designated representatives, do they have access to trade secret information contained within the PHA and other documents required to be developed by the standard? (Note that this access is subject to the provisions set forth in 1910.1200(i)(1).) FIELD NOTE REFERENCE(S):	.119(p)(3)	

Chapter 2 - PSM Regulatory Framework   43

2.7 STATE PROGRAMS

In its development of the PSM Standard, OSHA reviewed the provisions of state regulatory programs governing chemical process safety, most notably the programs in Delaware, California and New Jersey. (See Delaware Code, title 7, ch. 77, Extremely Hazardous Substances Risk Management Regulations; California Health and Safety Code, ch. 6.95, article 2, Hazardous Materials Management; New Jersey Administrative Code 7:31-1, 2, 3, 4; Toxic Catastrophe Prevention Act.) Companies operating facilities in those states must know and understand the specific state requirements, which are in many instances more stringent than the federal PSM regulations.

2.7.1 Delaware

Delaware's process safety management regulation was adopted in September 1989. Delaware was the first state to regulate flammable substances, and developed a unique method for determining the quantity of extremely hazardous substances (EHSs) that would have a catastrophic effect. The Delaware regulation requires implementation of a risk management program (RMP) and contains penalty provisions for noncompliance. The OSHA PSM Standard includes nine toxic substances not regulated by Delaware. Both Delaware and OSHA used the Substance Hazard Index (SHI) to rank the risk associated with each EHS. OSHA chose to regulate EHSs with a SHI of 5,000 or greater, whereas Delaware regulates EHSs with a SHI of 8,000 or greater, thus accounting for the difference.

The recently issued OSHA regulations closely parallel the Delaware RMP. The following is a comparison of the RMP elements for the Delaware and OSHA regulations:

OSHA	DELAWARE
Process Safety Information	Design Standards
Process Hazard Analysis	Process Hazard Analysis
Operating Procedures	Operating Instructions
Training	Operation Training
Contractors	Maintenance
Pre-Startup Review	Design Standards
Mechanical Integrity	Maintenance
Hot Work	Operating Instructions
Management of Change	Modification Control
Incident Investigation	Incident Investigation
Emergency Planning	Emergency Response
Compliance Safety Audits	Self Audit

The OSHA regulations require employee involvement in the development of the RMP, whereas the Delaware regulations do not. However, some Delaware facilities have used employees to assist in RMP development.

The Delaware program has been successful in reducing the probability of catastrophic incidents. Some facilities have replaced an EHS with a non-EHS material, while others reduced inventories below levels that could result in a catastrophic event. The biggest successes have been with facilities that have upgraded their procedures and practices to comply with the program (Barrish, 1993).

2.7.2 California

California is presently in the process of revising its chemical process safety regulations. Numerous industry representatives have complained that the proposed revisions to California's process safety standard (Title 8 of the California Code of Regulations, Section 5189) go too far and place an onerous training burden on employers subject to the standard. Among the provisions under attack, under Section 5189(d) of the proposal, employers would be required to develop and maintain a compilation of written safety information so employees can identify and understand hazards posed by processes involving acutely hazardous, flammable, and explosive materials. Industry representatives assert that this proposed revision presents massive employee communications obligations to employers.

California's current process safety standard applies only to refineries, chemical plants, and other manufacturing facilities in Standard Industrial Classification codes 28 and 29. The proposed revisions would broaden the scope of the standard to include all employers that handle 136 substances listed in an appendix as well as substances listed in the health and safety code that are incorporated by reference.

The current state standard was adopted in July 1992 pursuant to a 1990 state law and differs from the OSHA standard in several respects. In November 1992, OSHA determined California's standard is not "at least as effective as" the federal standard, triggering the proposed revisions to the state standard.

Because of concurrent jurisdiction, employers in California not subject to the state standard still must comply with the OSHA

46 OSHA and EPA PSM Requirements

rule. Employers outside of SIC codes 28 and 29 that use substances listed in the federal standard are subject to federal jurisdiction.

The proposal before the California Occupational Safety and Health Standards Board is designed to incorporate OSHA's requirements into the framework of the state standard while still complying with the 1990 state law.

2.8 OSHA COMPLIANCE DIRECTIVE

As can be expected following the issuance of any new regulatory standard, regulated companies are currently struggling to ensure compliance with the PSM Standard. To assist companies with compliance, OSHA has issued a regulatory compliance directive. OSHA Instruction CPL 2-2.45A was published by OSHA on September 28, 1992. OSHA expects to release a revision of this compliance directive some time in Fall 1994. For information concerning the status of the revised compliance directive, call OSHA's Publications Office at (202) 219-4667. A copy of OSHA Instruction CPL 2-2.45A is reproduced in **Appendix 9**. The compliance guidance provided in this book is based in part on the provisions of this important directive.

Industry officials raised concerns over the original 95-page instruction, which directed compliance officers to cite all violations of the process safety management standard at least as serious. The agency reasoned that any violations of the process safety standard could result in death or serious physical harm to employees. (See "Industry Representatives Restate Concerns Over OSHA Revision of Compliance Instruction," Occupational Safety & Health Daily (BNA) (Aug. 11, 1993).)

Union representatives questioned the effectiveness of the performance-based rule in protecting worker safety and health, concerns that led to a court challenge (*United Steelworkers v. OSHA*, 3d Cir., No. 92-3106, settlement announced 4/9/93). In settling the case, OSHA agreed to clarify its interpretation of the final standard by making several changes to its compliance instruction, which are to be included in the revised version.

2.9 OSHA INSPECTIONS AND ENFORCEMENT

As of October 1993, OSHA had conducted a total of 72 process safety management inspections since the chemical rule went into effect in May 1992, most of which were triggered by

Chapter 2 - PSM Regulatory Framework   47

employee complaints, fatality investigations, and referrals from other agencies. (See "Most Inspections Triggered by Complaints, Referrals, and Fatalities, OSHA Official Says," Occupational Safety & Health Daily (BNA) (Oct. 15, 1993).)

The relatively low number of inspections conducted thus far is an indication that an employer covered by the rule has only a small chance of receiving a process safety management inspection--unless its employees feel shut out of company safety programs or if hazards are evident to other agencies such as a state fire marshal.

As of October 1993, OSHA had proposed more than 220 citations for violations of the standard. About 90 percent of the citations were for violations classified at least as serious. That is, in part, the result of the agency's compliance instruction, issued in September 1992. The document, CPL 2-2.45A, instructs compliance officers to consider most alleged violations of the standard at least as serious, since any violations could result in death or serious harm to employees.

2.9.1 Inspection Totals

Employers can reduce the chances of an employee complaint by encouraging thorough employee participation in safety and health programs in the workplace, much of which is required under the process safety rule.

Nineteen programmed process safety management inspections were conducted between May and September 1992, 11 of which were targeted at high-hazard industries. Programmed inspections are planned in advance by the agency.

The agency's high-hazard list is culled from the Bureau of Labor Statistics (BLS) annual injury and illness survey, which provides a ranking of industries with relatively high rates of worker injuries and illnesses but no specific names of employers. OSHA does not have access to the data for individual firms, in part because BLS has pledged confidentiality to employers who participate in the survey. OSHA plans an additional 12 programmed process safety inspections in fiscal 1994.

Alternatively, unprogrammed inspections are typically triggered by a fatality, severe workplace injuries, or inspections by other agencies or by OSHA compliance officers conducting a general job safety and health inspection. The 53 unprogrammed

48 OSHA and EPA PSM Requirements

process safety inspections were triggered by employee complaints, fatalities and catastrophes, or referrals.

Of those 53 unprogrammed inspections, 19 were the result of employee complaints; eight were triggered by worker fatalities or catastrophes; 21 were a result of other agency referrals; and five were referrals from other OSHA compliance officers who had conducted a non-process safety management inspection.

2.9.2 PSM Provisions Cited

Of the 72 total inspections, chemical firms drew the most inspections, 26. Other industries inspected under the rule include food products (11 inspections), petroleum processors and manufacturers (nine), construction (nine), non-manufacturing, non-construction (nine), and other manufacturing (four).

By provision, OSHA proposed 65 citations against employers for allegedly violating 29 CFR 1910.119 (e), which requires the creating, updating, and implementation of process safety hazard analyses.

2.9.3 Violations of Operating Procedures

Employers who allegedly failed to develop and implement written operating procedures under the Occupational Safety and Health Administration's process safety management rule racked up the highest number of citations in the second year of the rule's implementation. (See "Lack of Written Operating Procedures Most Often Cited Under OSHA Rule, OSHA Says," Occupational Safety & Health Daily (BNA), Page d7 (Mar. 10, 1994).)

OSHA issued citations for 60 violations of the requirement between May 1, 1993, and Feb. 13, 1994. The provision, under 1910.119(f), requires covered chemical and petrochemical plants to develop procedures for safely operating each covered process, from initial startup to emergency shutdowns.

The data stemmed from inspections of 58 establishments nationwide over the nine-month period dating from May 1993, the first year anniversary of process safety management requirements that went into effect in May 1992. The 60 alleged violations of the operating procedures requirement is almost as many violations as inspections.

2.9.4 Specific Violations

For many of the citations, chemical process operators were generally well-versed in the plant's operating procedures but those procedures were never recorded or updated by the employer.

After the written program, the next most frequently cited standard sections were failure to:

*Provide initial employee training, 1910.119(g)(1); 24 violations;
*Develop a written plan for employee participation, 1910.119(c)(1); 22 violations;
*Develop and maintain written procedures to maintain the mechanical integrity of process equipment, 1910.119(j)(2); 22 violations;
*Provide an evaluation of a contractor's safety record, 1910.119(h)(2); 22 violations;
*Develop and implement procedures to ensure that process equipment is inspected and tested, 1910.119(j)(4); 21 violations; and
*Develop and maintain process safety information, 1910.119(d)(2); 19 violations.

Many of the alleged violations are the result of employers' failure to ensure that workers are able to participate in developing the various procedures required under the rule or to have access to those procedures.

Most of the citations allege serious violations of the process safety rule. Other OSHA officials have estimated that as much as 90 percent of the process safety-related citations have been issued for violations classified at least as serious, which is a direct result of an enforcement policy set under an agency compliance instruction (OSHA Instruction CPL 2-2.45A) issued in 1992 (23 OSHR 518).

2.9.5 $1.6 Million Fine Proposed for Violation of PSM Standard

A West Virginia fertilizer and pesticide manufacturer was hit with nearly $1.6 million in proposed penalties by the Occupational Safety and Health Administration for an August 1993 explosion that killed an employee and seriously injured two other workers. (See "OSHA Proposes $1.6 Million Fine, Citations in 1993 Blast at W. Va. Fertilizer Manufacturer," Occupational Safety & Health Daily (BNA) Page d2 (Feb. 18, 1994).)

OSHA issued citations alleging 23 willful and three serious violations of OSHA's process safety management standard, and one willful violation of the OSHA standard for hazardous waste and emergency response. The penalty and citations are consistent with the department's renewed emphasis on issuing high penalties when workers' lives are threatened.

The August 18, 1993 explosion and fire at Rhone-Poulenc AG Co.'s plant in Institute, West Virginia, triggered the release of more than 45,000 pounds of more than a dozen separate chemicals in a large cloud over the Kanawha River Valley. One worker died and two other workers suffered extreme damage to their lungs and skin as a result of exposure to the highly reactive chemicals released in the explosion.

2.9.6 Specific Citations

According to OSHA, the explosion and fire occurred in the chlorination reactor section of the plant, which produces a component used in the commercial pesticide Larvin. About 77 of the plant's 1,200 workers were employed in the chlorination unit.

OSHA issued the company instance-by-instance citations with a maximum penalty of $70,000 for each of the seven workers exposed during the chemical release - including the worker who died - for a total of $490,000. The company failed to develop and implement safe hazard controls as required under the process safety standard.

The seven workers were non-essential employees that were working in the area surrounding the chlorination reactor and were not evacuated during the release. The process safety standard requires employers to ensure that workers are trained to evacuate such areas once an emergency is declared, but the company allegedly did not signal an alarm once the incident began.

OSHA issues willful violations when it alleges a company has intentionally disregarded or operated with indifference to occupational safety and health laws and regulations.

2.9.7 Other Penalties

OSHA also issued penalties of $70,000 each for eight other alleged willful violations of the process safety standard for a total penalty of $560,000. Those fines were issued for alleged violations of various subparts of the standard that require the company to implement a maintenance program to ensure mechanical integrity of process safety equipment; inspect or conduct testing of its piping system; correct unit deficiencies; and investigate near-miss incidents.

OSHA also issued citations against the company for:

*Thirteen alleged willful violations with a proposed penalty of $35,000 each for allegedly failing to provide clear startup or emergency shutdown procedures and failing to address special or unique hazards in its written operating procedures; and failure to annually certify procedure accuracy. Those violations drew a total proposed penalty of $455,000;

*One alleged willful violation of requirements that the company address decontamination procedures in its emergency response plan, for a proposed penalty of $70,000; and

*Three serious violations at $7,000 per violation for allegedly failing to provide for employee participation in the plant's process safety management programs and for not giving workers access to process safety information.

2.9.8 Settlement of 1989 Phillips Petroleum Explosion

Phillips Petroleum Co. agreed March 2 to an out-of-court settlement with 191 individuals injured in a 1989 deadly explosion at its Pasadena, Texas, refinery (*Bogle v. Phillips,* Texas Dist. Ct., No. 89-46055, settlement reached 3/2/94; See also "Confidential Settlement Reached in Lawsuit from Phillips Explosion," Occupational Safety & Health Daily (BNA) page d4 (Mar. 10, 1994).)

Phillips reached the settlement three weeks after a jury trial began in which 15 victims were seeking damages in a federal district court in Texas. The amount of the settlement is confidential. Workers who settled suffered alleged physical and

52 OSHA and EPA PSM Requirements

psychological problems. Injuries included alleged back problems, hearing loss, and hernias.

The explosion occurred October 23, 1989, when air hoses that opened and shut a valve designed to keep gases inside a reactor loop and away from maintenance areas were reconnected too soon and the valve opened, allowing the unintentional release of explosive hydrocarbons. Twenty-three workers were killed; as many as 350 were injured.

The Occupational Safety and Health Administration originally proposed a $5.6 million fine against Phillips. The parties eventually settled on a $4 million fine. About 140 claims remain against Phillips. Of that number, 55 involve individuals who were not on the premises when the accident occurred; some lived in nearby residential areas. Phillips previously settled suits brought by families of explosion victims as well as injured workers.

The first case arising from the accident went to trial in 1993, and resulted in a $730,000 jury award for Phillips workers who claimed they suffered physical and mental injuries as a result of the explosion. The workers had sought $200 million. The jury found Phillips was negligent in its safety procedures at the plant, but exonerated the plant on claims of gross negligence (*McDaniels v. Phillips*, Texas Dist. Ct., Harris County, No. 90-34009, verdict 11/23/93).

Chapter 3

PSM Program Implementation and Operation

3.1 COMPONENTS OF A PSM SYSTEM

Process safety management is a system of management controls whereby process hazards are identified, understood, and controlled so that process related injuries and incidents are minimized or eliminated. Process safety management should be used for setting up lines of defense, auditing safety maintenance, and reducing risks to the lowest practical level. Components of PSM systems are described below. To successfully implement the PSM program, upper management support is needed, proper training must be initiated at the plant level, and detailed analysis of information on plant processes must be performed (Toth, 1990).

Numerous articles and books have been written about process safety management programs, many of which are listed in **Appendix 4**. The focus of this book is more specifically on compliance with the new PSM Standard. Still, PSM Standard compliance cannot be successfully acheived without a clear understanding of the components of a good process safety management system. These components are illustrated for the reader in the checklists that follow.

3.1.1 Generic Characteristics of a Management System[1]

Planning
[ ] Explicit Goals and Objectives
[ ] Well-defined Scope
[ ] Clear-cut Desired Outputs
[ ] Consideration of Alternative Achievement Mechanisms
[ ] Well-defined Inputs and Resource Requirements
[ ] Identification of Needed Tools and Training

Organizing
[ ] Strong Sponsorship
[ ] Clear Lines of Authority
[ ] Explicit Assignment of Roles and Responsibilities
[ ] Formal Procedures
[ ] Internal Coordination and Communication

Implementing
[ ] Detailed Work Plans
[ ] Specific Milestones for Accomplishments
[ ] Initiating Mechanisms
[ ] Execution
[ ] Resource Management

Controlling
[ ] Performance Standards and Measurement Methods
[ ] Checks and Balances
[ ] Performance Measurement and Reporting
[ ] Internal Reviews
[ ] Variance Procedure
[ ] Audit Mechanisms
[ ] Corrective Action Mechanisms
[ ] Procedure Renewal and Reauthorization

---

[1] Adapted from Appendix A to CCPS, *Guidelines for Technical Management of Process Safety*, AIChE 1989; *See also* Sweeney, 1992.

Chapter 3 - PSM Program Implementation   55

3.1.2 Components of a Process Safety Management Program[2]

Accountability: Objective and Goals
- [ ] Continuity of Operations
- [ ] Continuity of Systems (resources and funding)
- [ ] Continuity of Organizations
- [ ] Company Expectations
- [ ] Quality Process
- [ ] Control of Exceptions
- [ ] Alternative Methods
- [ ] Management Accessibility
- [ ] Communications

Process Knowledge and Documentation
- [ ] Process Definition and Design Criteria
- [ ] Process and Equipment Design
- [ ] Company Management of Information
- [ ] Documentation of Risk Management Decisions
- [ ] Protective Systems
- [ ] Normal and Upset Conditions
- [ ] Chemical and Occupational Health Hazards

Capital Project Review and Design Procedures
- [ ] Appropriation Request Procedures
- [ ] Risk Assessment for Investment Purposes
- [ ] Hazard Review
- [ ] Siting - Relative to Risk Management
- [ ] Pilot Plant
- [ ] Process Design and Review Procedures
- [ ] Project Management Procedures

Process Risk Management
- [ ] Hazard Identification
- [ ] Risk Assessment of Existing Operations
- [ ] Reduction of Risk
- [ ] In-Plant Emergency Response and Mitigation
- [ ] Process Management During Emergencies

---

[2] Adapted from CCPS, *Guidelines for Technical Management of Process Safety*, AIChE 1989; See also Gideon & Carmody, 1992.

56 OSHA and EPA PSM Requirements

[ ] Encouraging Client and Supplier Companies to Adopt Similar Risk Management Practices
[ ] Selection of Businesses With Acceptable Risks

Management of Change
[ ] Change of Technology
[ ] Change of Facility
[ ] Organizational Changes That May Impact Process Safety
[ ] Variance Procedures
[ ] Temporary Changes
[ ] Permanent Changes

Process and Equipment Integrity
[ ] Reliability Engineering
[ ] Materials of Construction
[ ] Fabrication and Inspection Procedures
[ ] Installation Procedures
[ ] Preventive Maintenance
[ ] Process, Hardware, and Systems Inspections and Testing (Pre-Startup Safety Review)
[ ] Maintenance Procedures
[ ] Alarm and Instrument Management
[ ] Demolition Procedures

Human Factors
[ ] Human Error Assessment
[ ] Operator/Process and Equipment Interfaces
[ ] Administrative Controls vs. Hardware

Training and Performance
[ ] Definition of Skills and Knowledge
[ ] Training Programs (e.g., new employees, contractors, technical employees)
[ ] Design of Operating and Maintenance Procedures
[ ] Initial Qualification Assessment
[ ] Ongoing Performance and Refresher Training
[ ] Instructor Program
[ ] Records Management

Incident Investigation
[ ] Major Incidents
[ ] Near-Miss Reporting

Chapter 3 - PSM Program Implementation   57

[ ]   Follow-up and Resolution
[ ]   Communication
[ ]   Incident Recording
[ ]   Third-Party Participation as Necessary

Standards, Codes, and Laws
[ ]   Internal Standards, Guidelines, and Practices (past history, flexible performance standards, amendments and upgrades)
[ ]   External Standards, Guidelines, and Practices

Audits and Corrective Actions
[ ]   Process Safety Audits and Compliance Reviews
[ ]   Resolutions and Close-Out Procedures

Enhancement of Process Safety Knowledge
[ ]   Internal and External Research
[ ]   Improved Predictive Systems
[ ]   Process Safety Reference Library

With a good PSM system in place, companies subject to the requirements of the PSM Standard are equipped to comply with the regulatory requirements of the OSHA PSM Standard. **Figure 3A** presents a useful flowchart of decisionmaking related to the overall process safety management system. An understanding of the PSM system is necessary to properly compile and analyze the process safety information and fulfill the other requirements of the PSM Standard.

58   OSHA and EPA PSM Requirements

Figure 3A - PSM Flowchart

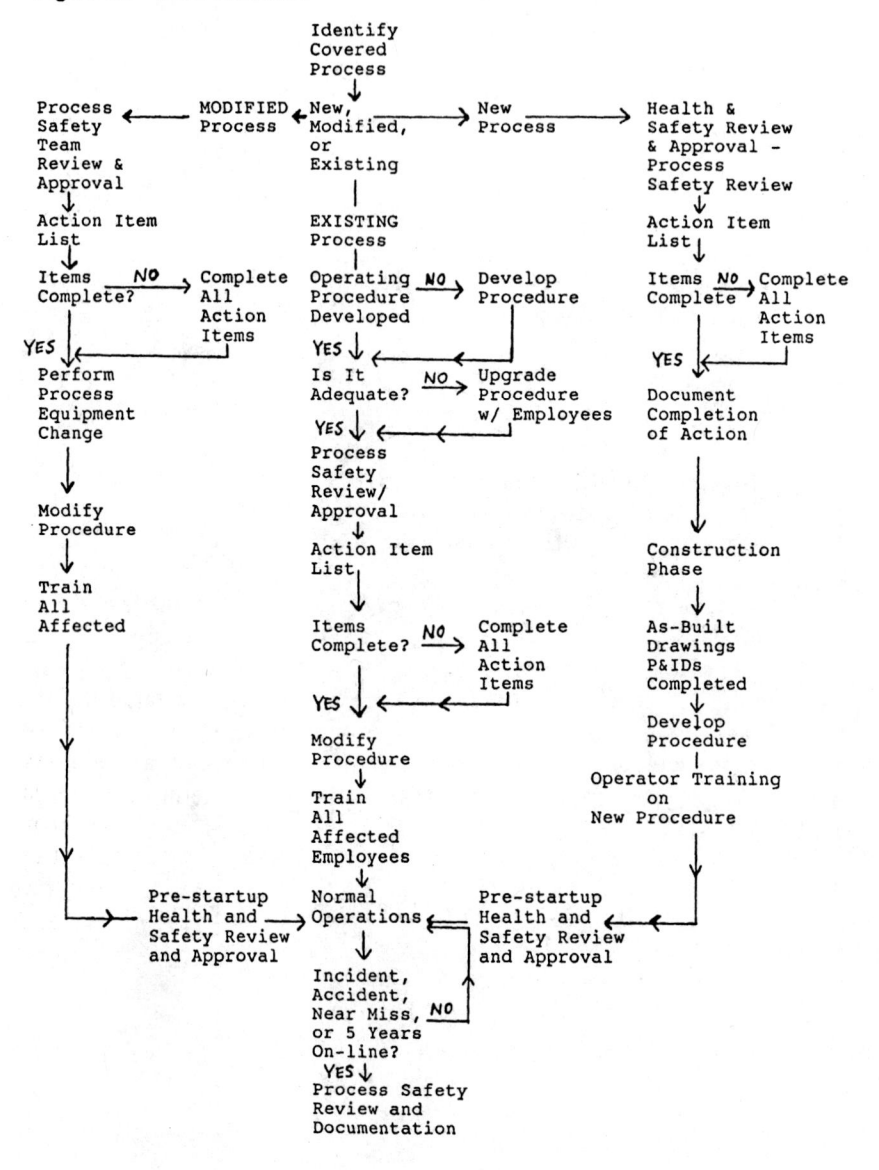

```
                              Identify
                              Covered
                              Process
                                 ↓
Process    ← MODIFIED ←  New,              →   New         →  Health &
Safety       Process     Modified,             Process        Safety Review
Team                     or                                   & Approval -
Review &                 Existing                             Process
Approval                    |                                 Safety Review
   ↓                        |                                     ↓
Action Item              EXISTING                             Action Item
List                     Process                              List |
   ↓                        |                                     ↓
Items    NO  → Complete  Operating  NO  →  Develop            Items   NO  → Complete
Complete?      All       Procedure         Procedure          Complete       All
               Action    Developed                                           Action
YES |←         Items     YES ↓  ←                             YES |←         Items
    ↓                    Is It      NO  →  Upgrade                ↓
Perform                  Adequate?         Procedure          Document
Process                                    w/ Employees       Completion
Equipment                YES ↓ ←  ←                           of Action
Change                   Process                                  |
   |                     Safety                                   |
   ↓                     Review/                                  ↓
Modify                   Approval                             Construction
Procedure                   ↓                                 Phase
   ↓                     Action Item                              ↓
Train                    List |                               As-Built
All                           ↓                               Drawings
Affected                 Items     NO  →   Complete           P&IDs
   |                     Complete?         All                Completed
   |                                       Action                 ↓
   |                     YES ↓ ←  ←        Items              Develop
   |                                                          Procedure
   |                     Modify                                   |
   |                     Procedure                            Operator Training
   ↓                         ↓                                on
   |                     Train                                New Procedure
   |                     All
   |                     Affected
   |                     Employees
   |                         ↓
   |       Pre-startup   Normal              Pre-startup
   └───→   Health and  → Operations ←  ←  ←  Health and  ←  ←
           Safety Review     ↓                Safety Review
           and Approval      |                and Approval
                         Incident,
                         Accident,
                         Near Miss,  NO
                         or 5 Years ──┐
                         On-line?     |
                         YES ↓        |
                         Process Safety
                         Review and
                         Documentation
```

3.1.3 Compiling Process Safety Information

The first step toward compliance with the PSM Standard requires collection and analysis of process safety information. The process safety information package describes the hazards of a chemical process or operation. It also helps personnel to identify and understand the hazards involved. A process safety information package generally consists of three components: the hazards-of-materials section includes all pertinent data on the hazardous characteristics of each chemical; equipment design information provides a description of key equipment design data; and process documentation provides the information that describes the path to safe operation.

A thorough understanding of the consequences of operation outside established parameters provides the basis for development of appropriate process controls, thus ensuring safer operation. Without this understanding, management will find it difficult to develop suitable process controls during the design stage and to maintain these controls during routine operations.

3.2 PROCESS SAFETY INFORMATION: SPECIFIC REQUIREMENTS

Employers are required to compile written process safety information before conducting mandatory Process Hazard Analyses (PHAs), which are discussed in Chapter 4. Compilation of this information is completed on the same schedule as required for completion of PHAs. The process safety information enables the employer and its employees to first identify and then understand the hazards posed by those processes involving highly hazardous chemicals. The three main components of the process safety information are described in the list below.

1. Information on the hazards of any highly hazardous chemicals used or produced by the process:
   [ ]   Toxicity;
   [ ]   Permissible exposure limits;
   [ ]   Physical data;
   [ ]   Reactivity data;
   [ ]   Corrosivity data; and
   [ ]   Thermal and chemical stability data, and hazardous effects of inadvertent mixing of different materials.

60 OSHA and EPA PSM Requirements

2. Information on the technology of the process:
[ ] Block flow diagram or simplified process flow diagram;
[ ] Process chemistry;
[ ] Maximum intended inventory;
[ ] Safe upper and lower limits of such items as temperatures, pressures, flows or compositions; and
[ ] Evaluation of the consequences of deviations, including those affecting employee health and safety.

3. Information on the equipment in the process:
[ ] Materials of construction;
[ ] Piping and instrument diagrams;
[ ] Electrical classification;
[ ] Relief system design and design basis;
[ ] Ventilation system design;
[ ] Design codes and standards employed;
[ ] Material and energy balances for processes built after May 26, 1992 and;
[ ] Safety systems.

The provisions concerning process safety information must meet the requirements in section 304(c)(1) of the CAAA. In this section OSHA must require employers to: "Develop and maintain written safety information identifying workplace chemical and process hazards, equipment used in the processes, and technology used in the processes."

Paragraph (d) of the PSM Standard contains the requirements regarding process safety information. OSHA proposed that the employer develop and maintain certain important information about a covered process such as information about the hazards and characteristics of the chemicals used, information about the process technology and how it works and information about the process equipment. This process safety information is to be communicated to employees involved in operating the process.

The compilation of information concerning process chemicals, technology and equipment provides the foundation for identifying and understanding the hazards involved in a process and is necessary in the development of a complete and thorough process hazard analysis, as well as other provisions in the final

Chapter 3 - PSM Program Implementation 61

rule including management of change, operating procedures, and incident investigations, etc.

OSHA decided to allow the compilation of process safety information to occur on a schedule consistent with the schedule for conducting process hazard analyses as described in final paragraph (e)(1), and discussed in full in Chapter 4. It is necessary to assemble the process safety information specified in the regulations in order to conduct an adequate process hazard analysis. Therefore it is reasonable to allow the collection and compilation of process safety information on a given process to be completed before a process hazard analysis on that process is begun, instead of requiring the compilation of all process safety information on all processes to be completed before any process hazard analyses are begun.

Paragraph (c) of the rule, employee participation, requires that employees and their representatives must have access to process hazard analysis reports and to all other information required to be developed under the PSM Standard. Also, process safety information pertinent to the employees job tasks is required to be communicated to operating employees in paragraph (g); to contract employees in paragraph (h); and to maintenance employees in paragraph (j).

The process safety information required by paragraph (d)(1) pertains to the hazards of the highly hazardous chemicals in the process and should include: toxicity information; permissible exposure limits; physical data; reactivity data, corrosivity data; thermal and chemical stability data; and the hazardous effects of inadvertent mixing of different materials that could foreseeably occur. Most of this information may already be available from the material safety data sheet (MSDS). MSDSs would be acceptable in meeting this requirement to the extent that the required information is available on the MSDS.

Under paragraph (d)(2), the employer must develop and maintain information pertaining to the technology of the process itself. Paragraph (d)(2)(i) specifies the required information, which includes the following: a block flow diagram or simplified process flow diagram; process chemistry; maximum intended inventory; safe upper and lower limits for such factors as temperatures, pressures, flows or compositions; and the consequences of any deviation in the process including those affecting the safety and health of employees.

62 OSHA and EPA PSM Requirements

OSHA recognized in paragraph (d)(2)(ii) that it might be difficult to obtain technical information for older existing processes. OSHA decided to allow an alternate method of obtaining the technical information only for those processes where such information does not exist. Paragraph (d)(2)(ii) reads as follows:
"Where the original technical information no longer exists, such information may be developed in conjunction with the process hazard analysis in sufficient detail to support the analysis."

In paragraph (d)(3)(i) OSHA requires that information be compiled concerning equipment used in the process including: materials of construction; piping and instrument diagrams (P&IDs); electrical classification; relief system design and design basis; ventilation system design; design codes employed; material and energy balances for processes built after the effective date of this standard; and safety systems (such as interlocks, detection, monitoring and suppression systems).

Paragraph (d)(3)(iii) requires that when existing equipment was designed and constructed in accordance with codes, standards, or practices that are no longer in general use, the employer must ascertain that the equipment is designed, installed, maintained, inspected, tested and operated in such a way that safe operation is assured.

There are many instances where process equipment has been in use for many years. Sometimes the codes and standards to which the equipment was initially designed and constructed are no longer in general use. For this type of situation, OSHA requires assurance that the older equipment still functions safely, and is still appropriate for its intended use. OSHA does not specify the method for this documentation. The employer is permitted to use any of several methods such as: documenting successful prior operation procedures; documenting that the equipment is consistent with the latest editions of codes and standards; or performing an engineering analysis to determine that the equipment is appropriate for its intended use.

3.2.1 Process Safety Information: Compliance Tips

Complete and accurate written information concerning process chemicals, process technology, and process equipment is essential to an effective process safety management program and to a process hazards analysis. The compiled information will be a

Chapter 3 - PSM Program Implementation 63

necessary resource to a variety of users including the team that will perform the process hazards analysis as required under paragraph (e); those developing the training programs and the operating procedures; contractors whose employees will be working with the process; those conducting the pre-startup reviews; local emergency preparedness planners; and incurance and enforcement officials.

The information to be compiled about the chemicals, including process intermediates, needs to be comprehensive enough for an accurate assessment of the fire and explosion characteristics, reactivity hazards, the safety and health hazards to workers, and the corrosion and erosion effects on the process equipment and monitoring tools. Current material safety data sheet (MSDS) information can be used to help meet this requirement which must be supplemented with process chemistry information including runaway reaction and over pressure hazards if applicable.

Process technology information will be a part of the process safety information package and it is expected that it will include diagrams of the type shown in **Appendix 3**, as well as employer established criteria for maximum inventory levels for process chemicals; limits beyond which would be considered upset conditions; and a qualitative estimate of the consequences or results of deviation that could occur if operating beyond the established process limits. Employers are encouraged to use diagrams which will help users understand the process.

A block flow diagram is used to show the major process equipment and interconnecting process flow lines and show flow rates, stream composition, temperatures, and pressures when necessary for clarity. The block flow diagram is a simplified diagram. Process flow diagrams are more complex and will show all main flow streams including valves to enhance the understanding of the process, as well as pressures and temperatures on all feed and product lines within all major vessels, in and out of headers and heat exchangers, and points of pressure and temperature control. Also, materials of construction information, pump capacities and pressure heads, compressor horsepower and vessel design pressures and temperatures are shown when necessary for clarity. In addition, major components of control loops are usually shown along with key utilities on process flow diagrams.

Piping and instrument diagrams (P&IDs) may be the more appropriate type of diagrams to show some of the above details and to display the information for the piping designer and engineering staff. The P&IDs are to be used to describe the relationships between equipment and instrumentation as well as other relevant information that will enhance clarity. Computer software programs which do P&IDs or other diagrams useful to the information package, may be used to help meet this requirement.

The information pertaining to process equipment design must be documented. In other words, what were the codes and standards relied on to establish good engineering practice. These codes and standards are published by such organizations as the American Society of Mechanical Engineers, American Petroleum Institute, American National Standards Institute, National Fire Protection Association, American Society for Testing and Materials, National Board of Boiler and Pressure Vessel Inspectors, National Association of Corrosion Engineers, American Society of Exchange Manufacturers Association, and model building code groups.

In addition, various engineering societies issue technical reports which impact process design. For example, the American Institute of Chemical Engineers has published technical reports on topics such as two phase flow for venting devices. This type of technically recognized report would constitute good engineering practice.

For existing equipment designed and constructed many years ago in accordance with the codes and standards available at that time and no longer in general use today, the employer must document which codes and standards were used and that the design and construction along with the testing, inspection and operation are still suitable for the intended use. Where the process technology requires a design which departs from the applicable codes and standards, the employer must document that the design and construction is suitable for the intended purpose.

3.2.2 Process Safety Information Compliance Checklist

1910.119 (d): PROCESS SAFETY INFORMATION

I. PROGRAM SUMMARY

The intent of this paragraph is to provide complete and accurate information concerning the process which is essential for an effective process safety management program and for conducting process hazard analyses. Therefore in accordance with the schedule set forth in paragraph (e)(1) the employer is required to compile written process safety information on process chemicals, process technology, and process equipment before conducting any process hazard analysis.

II. QUALITY CRITERIA REFERENCES

A.	1910.119(d):	Process Safety Information
B.	1910.119(e)(1):	Process Hazard Analysis
B.	1910.1200:	Hazard Communication

III. VERIFICATION OF PROGRAM ELEMENTS

	Criteria Reference	Met Y/N
A. Records Review 1. Has written process safety information been compiled before conducting any process hazard analysis (PHA)? FIELD NOTE REFERENCE(S):	.119(d)	

66 OSHA and EPA PSM Requirements

2. Is information included pertaining to the hazards of the highly hazardous chemicals used or produced by the process, and does the information include at least: • toxicity information • PEL's • physical data • reactivity data • corrosivity data • thermal and chemical stability data • hazardous effects of inadvertent mixing of different materials that could foreseeable occur? NOTE: MSDS's meeting the requirements of 29CFR1910.1200(g) may be used to the extent they contain the information required. FIELD NOTE REFERENCE(S):	.119(d)(1) .1200(g)	
3. Is information included concerning the technology of the process, and does it include at least: • a block flow diagram or simplified process flow diagram? • process chemistry? • maximum intended inventory? • safe upper and lower limits? • an evaluation of the consequences of deviations? (Where the original technical information no longer exists, it may be developed in conjunction with the PHA.) FIELD NOTE REFERENCE(S):	.119(d)(2)	
4. Is information included pertaining to equipment in the process, and does it include at least: • materials of construction? • piping and instrument diagrams (P&ID's)? • electrical classification? • relief system design and design basis? • ventilation system design? • design codes and standards employed? • material and energy balances for processes built after May 26, 1992? • safety systems (e.g. interlocks, detection or suppressions systems)? FIELD NOTE REFERENCE(S):	.119(d)(3)(i)	

Chapter 3 -- PSM Program Implementation

5. Has the employer documented that equipment complies with recognized, generally accepted good engineering practices? (Review the documentation for evidence that compliance with the appropriate consensus standards has been researched.) FIELD NOTE REFERENCE(S):	.119(d)(3)(ii)
6. Has the employer determined and documented that existing equipment designed and constructed in accordance with codes, standards, or practices no longer in general use are designed, maintained, inspected, tested, and operating in a safe manner? (Documentation may be through methods such as: documenting successful prior operation procedures; documenting that the equipment is consistent with the appropriate editions of codes and standards; or performing an engineering analysis to determine that the equipment is appropriate for its intended use.) FIELD NOTE REFERENCE(S):	.119(d)(3)(iii)

B. On-site Conditions

1. Do observations of a representative sample of process chemicals and equipment indicate that the process information is complete? (Information that does not correspond to the actual conditions demonstrates incomplete information. Check critical equipment and components to see if they have been properly identified.) FIELD NOTE REFERENCE(S):	.119(d)
2. Do observations of a representative sample of process components indicate that the process complies with recognized and generally accepted good engineering practice? (Review a representative number of safety devices such as pressure relief devices for proper sizing according to the maximum anticipated pressure.) FIELD NOTE REFERENCE(S):	.119(d)(3)(ii)

68 OSHA and EPA PSM Requirements

3. Do observations of a representative sample of the existing equipment designed and constructed according to codes, standards, or practices no longer in general use indicate that this equipment is inspected and is operated in a safe manner (as documented by the employer)? FIELD NOTE REFERENCE(S):	.119(d) (3) (iii)
C. Interviews Process Hazard Analysis (PHA) Team: 1. Based on interviews with a representative number of PHA team members, was the process safety information complete before the process hazard analysis was conducted? FIELD NOTE REFERENCE(S):	.119(d)
Operators: 2. Based on interviews with a representative number of operators, is MSDS information readily available to the operators who work with hazardous materials? FIELD NOTE REFERENCE(S):	.1200
Engineers (if any; or other qualified persons capable of providing the information requested; see NOTE, p. A-2): 3. Based on interviews with a representative number of engineers, has the employer documented that the process equipment complies with recognized and generally accepted good engineering practice? (Ask about the technical bases for design and selection of equipment, the materials of construction, electrical classifications, relief devices sizing versus maximum anticipated pressures, installation procedures to assure equipment meets design specifications, etc.) FIELD NOTE REFERENCE(S):	.119(d) (3)(ii)

3.3 EMPLOYEE INVOLVEMENT IN PROCESS SAFETY MANAGEMENT

Section 304 of the Clean Air Act Amendments states that employers are to consult with their employees and their representatives regarding the employers efforts in the development and implementation of the process safety management program elements and hazard assessments. Section 304 also requires employers to train and educate their employees and to inform affected employees of the findings from incident investigations required by the process safety management program. Many employers, under their safety and health programs, have already established means and methods to keep employees and their representatives informed about safety and health issues and employers may be able to adapt these practices and procedures to meet their obligations under this standard. Employers who have not implemented an occupational safety and health program may wish to form a safety and health committee of employees and management representatives to help the employer meet the obligations specified by this standard. These committees can become a significant ally in helping the employer to implement and maintain an effective process safety management program for all employees.

3.4 EMPLOYEE PARTICIPATION: SPECIFIC REQUIREMENTS

The issue of employee participation in process safety management received even greater attention after the Clean Air Act Amendments (CAAA) were signed. The CAAA contains a requirement in section 304(c)(3) that the employer "consult with employees and their representatives on the development and conduct of hazard assessments and the development of chemical accident prevention plans and provide access to these and other records required under the standard."

After a thorough analysis of the CAAA and the rulemaking on the PSM Standard, OSHA concluded that it is important for one member of each team be an employee who is knowledgeable about the process. This employee may very well be an employee representative or an employee representative may be participating on a team because of some expertise that the individual can contribute to the team. However, OSHA's view is

70 OSHA and EPA PSM Requirements

that it is not necessary or appropriate to mandate team membership on the basis of organization affiliation (i.e., union membership).

OSHA felt that the intent of the CAAA demands a broader approach to employee participation. Thus, OSHA requires that employers consult with employees and their representatives on the development and conduct of hazard assessments (OSHA's process hazard analyses) and consult with employees on the development of chemical accident prevention plans. Also, as prescribed by the CAAA, OSHA is requiring that all process hazard analyses and all other information required to be developed by the PSM Standard be available to employees and their representatives.

OSHA added language contained in the CAAA to the final PSM regulations as paragraph (c). This provision, which requires broad and active employee participation in all elements of the process safety management program through consultation, is intended to enhance the overall program. OSHA also believes that the CAAA requirements demand that an employer carefully consider and structure the plant's approach to employee involvement in the process safety management program. Consequently, the employer must address this issue to ensure that the employer actively considers the appropriate method of employee participation in the implementation of the process safety management program at the workplace. Thus, OSHA also included a specific requirement that an employer develop a plan of action on how the employer is going to implement the employee participation requirements.

Paragraph (c) of the PSM Standard reads as follows:
"Employee participation. Employers shall develop a written plan of action regarding the implementation of the employee participation required by this paragraph. Employers shall consult with employees and their representatives on the conduct and development of process hazard analyses and on the development of the other elements of process safety management in this standard. Employers shall provide to employees and their representatives access to process hazard analyses and to all other information required to be developed under this standard."

3.4.1 Employee Participation Compliance Checklist

1910.119 (c): EMPLOYEE PARTICIPATION

I. PROGRAM SUMMARY

The intent of this paragraph is to require employers to involve employees at an elemental level of the PSM program. Minimum requirements for an Employee Participation Program for PSM must include a written plan of action for implementing employee consultation on the development of process hazard analyses and other elements of process hazard management contained within 1910.119. The employer must also provide ready access to all the information required to be developed under the standard.

II. QUALITY CRITERIA REFERENCES

A. 1910.119(c): Employee Participation

III. VERIFICATION OF PROGRAM ELEMENTS

	Criteria Reference	Met Y/N
A. Records Review.		
1. Does a written program exist regarding employee participation? FIELD NOTE REFERENCE(S):	.119(c)(1)	
2. Does the written program include consultation with employees and their representatives on the conduct and development of process hazard analyses and on the development of other elements in the PSM standard? FIELD NOTE REFERENCE(S):	.119(c)(2)	
3. Does the written program provide employees and their representatives access to process hazard analyses and all other information developed as required by the PSM standard? FIELD NOTE REFERENCE(S):	.119(c)(3)	
B. On-site Conditions Not applicable.		

72 OSHA and EPA PSM Requirements

C. Interviews 1. Based on interviews with a representative number of employees and their representatives, have they been consulted on the conduct and development of the process hazard analyses? FIELD NOTE REFERENCE(S):	.119(c)(2)
2. Based on interviews with a representative number of employees and their representatives, have they been consulted on the development of other elements of the Process Safety Management program? FIELD NOTE REFERENCE(S):	.119(c)(2)
3. Based on interviews with a representative number of employees and their representatives, have they been provided access to process hazard analyses and to all other information required to be developed by the PSM standard? (Ask about unreasonable delays in access to information and whether time is given during the working hours to access information required by the PSM standard.) FIELD NOTE REFERENCE(S):	.119(c)(3)

Chapter 3 - PSM Program Implementation   73

3.5   PROCESS SAFETY OPERATING PROCEDURES

Operating procedures provide a clear definition of the safe path from raw material to finished product, and of the consequences of deviating from the documented path. Operating procedures should employ a clear, straightforward organizational format. They must include a comprehensive safety, occupational health and environmental control section. By taking advantage of the expertise of the people operating the process equipment, management can ensure the procedures and their accompanying sketches, tabulations, and graphs remain up-to-date.

For operating procedures to be effective, they must remain current. Once an operating procedure is firmly established and personnel are trained so that procedure and field practice are consistent, no change in technology, facilities or procedures can be permitted. That is unless the operating procedure is properly updated and authorized, and the personnel trained.

3.5.1   Written Operating Procedures

In conjunction with compilation of the process safety information, the regulations also require that employers implement written operating procedures that provide clear instructions for safely conducting activities concerning the covered process. Operating procedures must be readily available to all employees who work in or maintain a process. The operating procedures must be continuously reviewed and updated. The employer must certify annually that the operating procedures are current and accurate. The minimum components of an acceptable operating procedure for a process are outlined in the following checklist:

1. Steps for each operating phase:
   [ ]   Initial startup;
   [ ]   Normal operations;
   [ ]   Temporary operations;
   [ ]   Emergency shutdown, including the conditions under which emergency shutdown is required, and the assignment of shutdown responsibility to qualified operators;
   [ ]   Emergency operations;
   [ ]   Normal shutdown; and
   [ ]   Startup following a turnaround, or after an emergency shutdown.

2. Operating limits:
   [ ]   Consequences of deviation; and
   [ ]   Steps required to correct or avoid deviation.

3. Safety and health considerations:
   [ ]   Properties of, and hazards presented by, the chemicals used in the process;
   [ ]   Precautions necessary to prevent exposure, including engineering controls, administrative controls, and personal protective equipment;
   [ ]   Control measures to be taken if physical contact or airborne exposure occurs;
   [ ]   Quality control for raw materials and control of hazardous chemical inventory levels; and
   [ ]   Safety systems and their functions.

   The employer must develop and implement safe work practices to provide for the control of hazards during work activities such as lockout/tagout, confined space entry, opening process equipment or piping, and control over entrance into a facility by maintenance, contractor, laboratory, or other support personnel.

3.6 OPERATING PROCEDURES: SPECIFIC REQUIREMENTS

The provisions concerning operating procedures included in the PSM Standard must meet the requirements of sections 304(c) (6) and (7) of the CAAA which state that the OSHA standard must require employers to:

(6) Develop and implement written operating procedures for the chemical process including procedures for each operating phase, operating limitations, and safety and health considerations.

(7) Provide written safety and operating information to employees and training employees in operating procedures, emphasizing hazards and safe practices.

Chapter 3 - PSM Program Implementation   75

Paragraph (f) of the PSM Standard contains provisions requiring the development and implementation of written operating procedures. The procedures are to provide clear instructions for safely conducting activities involved in covered processes and they must be consistent with the process safety information. To have an effective process safety management program, tasks and procedures directly and indirectly related to the covered process must be appropriate, clear, consistent, and most importantly, communicated to employees.

Many different tasks may be necessary during a process, such as initial startup, handling special hazards, normal operation, temporary operations and emergency shutdown. The appropriate and consistent manner in which the employer expects these tasks and procedures to be performed is sometimes referred to as standard operating procedures.

It is important to have written operating procedures so employees working on a process do a given task in the same manner. There is less likelihood that incidents will occur if written operating procedures are developed so even a new employee or one who is relatively inexperienced will respond to a given event in a preconsidered and prescribed manner. It is also important that the procedures be written so that they can be communicated to employees in the most effective manner possible. Such written procedures comprise the employer's policy with respect to what is to be accomplished, and how it is to be accomplished safely. This will ensure that employees will perform like tasks and procedures in a consistently safe manner, and employees will know what is expected of them. These procedures must also be available for ready reference and review during production to make sure the process is operated properly.

In paragraph (f)(1)(i), OSHA requires that the operating procedures address steps for each operating phase, including initial startup, normal operation, temporary operations, emergency operations, normal shutdown, and startup following turnaround or emergency shutdown. In paragraph (f)(1)(ii) OSHA requires that the operating procedures address the process operating limits, including the following: consequences of deviation; steps required to correct and/or avoid deviation; and safety systems (including detection and monitoring equipment) and their functions.

In paragraph (f)(1)(iii), OSHA requires that the operating procedures address safety and health considerations regarding the process, including the following: properties of, and hazards presented, by the chemicals used; precautions necessary to prevent exposure; control measures to be taken if physical contact or airborne exposure occurs; safety procedures for opening process equipment (such as pipeline breaking); quality control for raw materials and control of hazardous chemicals inventory levels; and any special or unique hazards.

Proposed paragraph (f)(2) requires that a copy of the operating procedures be readily accessible to employees who work in or maintain a process. This requirement assures that a ready and up-to-date reference is available to employees when needed. It will also form a foundation for the employee training requirement discussed in **Chapter 5**. Operating procedures must be reviewed to ensure that they reflect current operating practices and any changes to the process or facility. Since it is extremely important to the safe operation of covered processes that operating procedures remain current and accurate, OSHA has added a precaution to guard against the use of outdated or inaccurate operating procedures by requiring that an employer verify annually that the operating procedures are current and accurate.

Paragraph (f)(4) requires that the employer develop and implement safe work practices to provide for the control of hazards during work activities. The objectives of these requirements are, first, to ensure that those persons operating high hazard processes are cognizant of any non-routine work (i.e., maintenance, construction, sampling or other activity) that is occurring in the process. The second objective is to ensure that those in responsible control of the facility are also in control of such non-routine work so as to ensure that the work does not undermine the safe control of the process. The third objective is to provide information to those workers performing non-routine work regarding the hazards and necessary precautions attendant to that work. Ordinarily, in chemical plants, maintenance and construction activities are supervised by persons other than those in direct control of the process. Implementation of these practices will ensure that control over all activity in high hazard plants remains with those who manage the production units while they are in operation. OSHA agrees that this approach will provide significant safety to employees impacted by on-going work activities and prefers this performance oriented approach provision.

Chapter 3 - PSM Program Implementation   77

3.6.1 Operating Procedures and Practices

Operating procedures describe tasks to be performed, data to be recorded, operating conditions to be maintained, samples to be collected, and safety and health precautions to be taken. The procedures need to be technically accurate, understandable to employees, and revised periodically to ensure that they reflect current operations. The process safety information package is to be used as a resource to better assure that the operating procedures and practices are consistent with the known hazards of the chemicals in the process and that the operating parameters are accurate. Operating procedures should be reviewed by engineering staff and operating personnel to ensure that they are accurate and provide practical instructions on how to actually carry out job duties safely.

Operating procedures will include specific instructions or details on what steps are to be taken or followed in carrying out the stated procedures. These operating instructions for each procedure should include the applicable safety precautions and should contain appropriate information on safety implications. For example, the operating procedures addressing operating parameters will contain operating instructions about pressure limits, temperature ranges, flow rates, what to do when an upset condition occurs, what alarms and instruments are pertinent if an upset condition occurs, and other subjects. Another example of using operating instructions to properly implement operating procedures is in starting up or shutting down the process. In these cases, different parameters will be required from those of normal operation. These operating instructions need to clearly indicate the distinctions between startup and normal operations such as the appropriate allowances for heating up a unit to reach the normal operating parameters. Also the operating instructions need to describe the proper method for increasing the temperature of the unit until the normal operating temperature parameters are achieved.

Computerized process control systems add complexity to operating instructions. These operating instructions need to describe the logic of the software as well as the relationship between the equipment and the control system; otherwise, it may not be apparent to the operator. Operating procedures and instructions are important for training operating personnel. The operating procedures are often viewed as the standard operating

78 OSHA and EPA PSM Requirements

practices (SOPs) for operations. Control room personnel and operating staff, in general, need to have a full understanding of operating procedures. If workers are not fluent in English then procedures and instructions need to be prepared in a second language understood by the workers. In addition, operating procedures need to be changed when there is a change in the process as a result of the management of change procedures. The consequences of operating procedure changes need to be fully evaluated and the information conveyed to the personnel. For example, mechanical changes to the process made by the maintenance department (like changing a valve from steel to brass or other subtle changes) need to be evaluated to determine if operating procedures and practices also need to be changed. All management of change actions must be coordinated and integrated with current operating procedures and operating personnel must be oriented to the changes in procedures before the change is made. When the process is shut down in order to make a change, then the operating procedures must be updated before startup of the process.

Training in how to handle upset conditions must be accomplished as well as what operating personnel are to do in emergencies such as when a pump seal fails or a pipeline ruptures. Communication between operating personnel and workers performing work within the process area, such as nonroutine tasks, also must be maintained.

Chapter 3 - PSM Program Implementation 79

3.6.2 Operating Procedures Compliance Checklist

1910.119 (f): OPERATING PROCEDURES

I. PROGRAM SUMMARY

The intent of this paragraph is to provide clear instruction for conducting activities involved in covered processes that are consistent with the process safety information. The operating procedures must address steps for each operating phase, operating limits, safety and health considerations, and safety systems and their functions.

II. QUALITY CRITERIA REFERENCES

A.	1910.119(f)	D.	1910.1000
B.	1910.120	E.	1910.1200
C.	1910.147		

III. VERIFICATION OF PROGRAM ELEMENTS

	Criteria Reference	Met Y/N
A. Records Review 1. Do written operating procedures exist for each covered process? Do the procedures provide clear instructions for conducting activities safely? FIELD NOTE REFERENCE(S):	.119(f) (1)	

2. Do the operating instructions address, as a minimum, steps for each operating phase, including:
• Initial start-up?
• Normal operations?
• Temporary operations?
• Emergency shutdowns?
  • Conditions requiring emergency shutdown?
  • Assignment of shutdown responsibility to qualified operators?
• Emergency operations?
• Normal shutdown?
• Start-ups following a turnaround or emergency shutdown?
 	
FIELD NOTE REFERENCE(S):
	.119(f)
	(1)(i)	

80 OSHA and EPA PSM Requirements

3. Do the operating procedures include operating limits that outline consequences of process deviation and steps required to correct or avoid deviations? FIELD NOTE REFERENCE(S):	.119(f) (1)(ii)
4. Have safety and health considerations been included in the operating procedures? Do they include at a minimum: • Properties of, and hazards presented by, chemicals used in the process? • Precautions necessary to prevent exposure, including engineering controls, administrative controls, and personal protective equipment? • Control measures to be taken if physical contact or airborne exposure occurs? • Quality control for raw materials and control of hazardous chemical inventory levels? • Any special or unique hazards? FIELD NOTE REFERENCE(S):	.119(f) (1) (iii)
5. Are safety systems and their functions included in the operating procedures? FIELD NOTE REFERENCE(S):	.119(f) (1)(iv)
6. Are the operating instructions consistent with the process safety information? FIELD NOTE REFERENCE(S):	.119(f) (1)
7. Are operating procedures readily accessible to employees who work in or maintain a process? FIELD NOTE REFERENCE(S):	.119(f) (2)

Chapter 3 - PSM Program Implementation   81

8. Are operating procedures reviewed as often as necessary to assure that they reflect current operating practice? Are they certified annually by the employer that they are current and accurate? Do they reflect current operating practice that have resulted from changes in: • Process chemicals? • Technology? • Equipment? • Facilities? FIELD NOTE REFERENCE(S):	.119(f) (3)
9. Have safe work practices been developed and implemented for employees and contractors to control hazards during operations such as: • Lockout/tagout? • Confined space entry? • Opening process equipment or piping? • Control over entrance into a facility by maintenance, contractor, laboratory or other support personnel? FIELD NOTE REFERENCE(S):	.119(f) (4)
B. On-site Conditions 1. Does observation of a representative sample of processes indicate that the written operating procedures are being implemented? FIELD NOTE REFERENCE(S):	.119(f) (1)
2. Does observation of a representative sample of processes indicate that the written operating procedures are readily accessible to employees who work or maintain a process? FIELD NOTE REFERENCE(S):	.119(f) (2)
3. Does observation of a representative sample of processes indicate that operating procedures reflect current practice, including changes that result from process chemicals, technology, equipment, and facilities? (Observe to see if actual procedures match the written operating procedures.) FIELD NOTE REFERENCE(S):	.119(f) (3)

4. Does observation of representative operations indicate that safe work practices have been implemented for company and contractor employees? Do such work practices include, where appropriate: • Lockout/tagout? • Confined space entry? • Opening process equipment or piping? • Control over entrance into a facility by maintenance, contractor, laboratory, and other support personnel? FIELD NOTE REFERENCE(S):	.119(f) (4)	

C. Interviews

1. Based on interviews with a representative number of operators, are the written operating procedures implemented for each covered process? FIELD NOTE REFERENCE(S):	.119(f) (1)	
2. Based on interviews with a representative number of operators, do operating procedures provide clear instructions for safely conducting activities? (Specifically ask for conditions requiring emergency shutdown, the operating limits of a particular process or item of equipment, what might occur if a deviation from those limits should take place, steps to avoid the deviation, and precautions necessary to prevent exposure to hazardous chemicals.) FIELD NOTE REFERENCE(S):	.119(f) (1)	
3. Based on interviews with a representative number of employees who work in or maintain a process, are the operating procedures readily accessible? FIELD NOTE REFERENCE(S):	.119(f) (2)	
4. Based on interviews with a representative number of operators and maintenance employees, do the operating procedures reflect current operating practice? FIELD NOTE REFERENCE(S):	.119(f) (3)	

Chapter 4

Conducting the Process Hazard Analysis

4.1 PHA METHODOLOGIES

The cornerstone of OSHA's PSM Standard is the Process Hazard Analysis (PHA). Process hazards analysis includes methods that effectively identify, evaluate and control hazards associated with process facilities. As another tool to prevent injuries and incidents, process hazards analysis uses orderly, organized, methodical approaches, and documents results for use in follow-up and training of personnel. Risk factors for hazardous chemicals include the process, the materials used, the installation and operation of the equipment, and the environment surrounding the process.

Often process hazards analysis is equated with a Hazard and Operability Study (HAZOP). This is, however, just one technique for performing a PHA. For example, when the "What-If" checklist is used with Failure Mode and Effect Analysis (FMEA) or HAZOP, the two methods complement each other and produce a higher quality result.

The PSM rule, being a performance-oriented standard, does not specify which methods must be used to complete PHAs. Regulated facilities are given the discretion to chose one method or a combination of methods for completion of PHAs mandated under paragraph (e) of the PSM Regulations.

Following completion of the process safety information phase described in Chapter 3, the process hazard analysis (PHA) must be completed to comply with the PSM Standard. A PHA is a thorough, orderly and systematic approach to identifying, evaluating and controlling processes involving highly hazardous chemicals. Employers must determine and document the priority order for conducting the PHAs. Priority order is determined by such considerations as the extent of the process hazards, the number of potentially affected employees, the age of the process, and the operating history of the process.

The following list summarizes various methods that the employer may use to determine and evaluate the hazards of the process being analyzed:

1. What-If. For relatively uncomplicated processes, the process is reviewed from raw materials to product. At each handling or processing step, "what if" questions are formulated and answered, to evaluate the effects of component failures or procedural errors on the process.

2. Checklist. For more complex processes, the "what if" study can be best organized through the use of a "checklist," and assigning certain aspects of the process to those members of the PHA team having the greatest experience or skill in evaluating those aspects. Operator practices and job knowledge are audited in the field, the suitability of equipment and materials of construction is studied, the chemistry of the process and the control systems are reviewed, and the operating and maintenance records are audited. Generally, a checklist evaluation of a process precedes use of the more sophisticated methods unless the process has been operated safely for many years and has been subjected to periodic and thorough safety inspections and audits.

3. What-If/Checklist. The what-if/checklist is a broadly based hazard assessment technique that combines the creative thinking of a selected team of specialists with the methodical focus of a prepared checklist. The result is a comprehensive hazard analysis that is extremely useful in training operating personnel on the hazards of the particular operation. The review team is selected to represent a wide range of disciplines, including production, mechanical, technical, and safety. Each person is given a basic information package regarding the operation to be studied. This package typically includes information on hazards of materials, process technology, procedures, equipment design, instrumentation control, incident experience, and previous hazard reviews. A field tour of the operation is also conducted at this time. The review team methodically examines the operation from receipt of raw materials to delivery of the finished product to the customer's site.

Chapter 4 - Process Hazard Analysis   85

At each step the group collectively generates a listing of "what-if" questions regarding the hazards and safety of the operation. When the review team has completed listing its spontaneously generated questions, it systematically goes through a prepared checklist to stimulate additional questions. Subsequently, answers are developed for each question. The review team then works to achieve a consensus on each question and answer. From these answers, a listing of recommendations is developed specifying the need for additional action or study. The recommendations, along with the list of questions and answers, become the key elements of the hazard assessment report.

4. Hazard and Operability Study (HAZOP). HAZOP is a formally structured method of systematically investigating each element of a system for all of the ways in which important parameters can deviate from the intended design conditions to create hazards and operability problems. The hazard and operability problems are typically determined by a study of the piping and instrument diagrams (or plant model) by a team of personnel who critically analyze effects of potential problems arising in each pipeline and each vessel of the operation. Pertinent parameters are selected, such as flow, temperature, pressure, and time. Then the effect of deviations from design conditions of each parameter is examined. A list of key words, such as "more of," "less of," "part of," are selected for use in describing each potential deviation. The system is evaluated as designed and with deviations noted. All causes of failure as well as any existing safeguards and protection are identified. An assessment is made by weighing the consequences, causes, and protection requirements involved.

5. Failure Mode and Effect Analysis (FMEA). The FMEA is a methodical study of component failures. This review starts with a diagram of the operation, and includes all components that could fail and conceivably affect the safety of the operation. Typical examples are instrument transmitters, controllers, valves, pumps, and rotometers. These components are listed on a data tabulation sheet and individually analyzed for the following:

*Potential mode of failure, i.e., open, closed, on, off, leaks, etc.;
*Consequence of the failure, i.e., effect on other components and effect on whole system;

86 OSHA and EPA PSM Requirements

*Hazard class, i.e., high, moderate, low;
*Probability of failure;
*Detection methods;
*Compensating provision/remarks.

Multiple concurrent failures are also included in the analysis. The last step in the analysis is to analyze the data for each component or multiple component failure and develop a series of recommendations appropriate to risk management.

6. Fault Tree Analysis. A fault tree analysis can be either a qualitative or a quantitative model of all the undesirable outcomes, such as a toxic gas release or explosion, which could result frcm a specific initiating event. It begins with a graphic representation (using logic symbols) of all possible sequences of events that could result in an incident. The resulting diagram looks like a tree with many branches, each branch listing the sequential events (failures) for different independent paths to the top event. Probabilities (using failure rate data) are assigned to each event and then used to calculate the probability of occurrence of the undesired event. This technique is particularly useful in evaluating the effect of alternative actions on reducing the probability of occurrence of the undesired event.

OSHA recommends that the PHA be performed by a team with expertise in engineering and process operations, and that the team should include at least one employee who has experience with and knowledge of the process being evaluated. One member of the team must also be knowledgeable of the specific analysis methods being used.

At least every five years after the initial completion of a PHA, the process hazard analysis must be updated and revalidated to ensure that the hazard analysis is consistent with the current process. Records of all PHAs, including updates and revalidations must be retained for the life of the process. The information must be stored such that it is easily accessible to all employees.

4.2 SELECTING THE APPROPRIATE METHODOLOGY

The selection of a PHA methodology or technique will be influenced by several factors, including the size and complexity of the process, as well as how much is known about the process. The application of a PHA to a process may involve the use of different methodologies for various parts of a process. For example, a PHA checklist may be used for a standard boiler or heater exchanger, and a PHA Hazard and Operability (HAZOP) study may be used for the overall process. Also, when an employer has a large continuous process with several control rooms for different portions of the process, such as a distillation tower and a blending operation, the employer may wish to do each segment separately and then integrate the final results.

The "What-If" checklist provides comprehensive coverage of a broad range of hazards, is relatively easy to use and is an effective learning tool. It can provide a complete review for a relatively uncomplicated process. However, the quality of this method depends on the experience and background of the reviewer, who must ask the right questions to perform an effective analysis.

FMEA provides a methodical approach to failure modes and consequences. FMEA works best when studying a specific item of equipment. Its semi-quantitative approach assists in prioritizing hazards. A drawback of the FMEA is that it assumes normal operation is satisfactory and requires an accurate model or diagram to proceed effectively.

HAZOP provides a methodical assessment of all deviations from normal operations in discovering how deviations from the intention of the design can occur, and in determining if the consequences of such deviations are hazardous. The limitation of HAZOP is that it assumes the design is correct for normal situations. It also requires a good team leader and a good model or diagram.

Fault tree analysis defines various routes to a top event, quantifies probability of reaching the top event, and provides objective information for decisionmaking. Fault tree analysis focuses on events rather than the process, and requires quantitative techniques and expertise. because of these limitations, a fault tree analysis is reserved for use in critical hazard situations.

4.3   PHA BENEFITS FOR SMALL BUSINESSES

There is a large number of smaller distributors and consumers of hazardous materials, which would not ordinarily be subject to the requirements of OSHA's PSM Standard because they

88 OSHA and EPA PSM Requirements

do not use extremely hazardous substances in the threshold quantities that trigger application of the standard. Still, these facilities can benefit from an understanding of the rigorous demands of the PSM Standard by implementing some of the standard's more general process safety management requirements and procedures (McKelvey et. al., 1992). Those facilities that are left more or less unregulated by the PSM Standard must nonetheless comply with other less stringent health and safety regulations and, therefore, should seek to implement many of the methodologies for accident prevention, process hazard analysis and employee training set forth by the PSM Standard. Examples of operations that may employ more generalized PSM techniques include: use of chlorine gas in municipal water treatment systems; ammonia-based commercial refrigeration systems; ethylene oxide sterilizers in hospitals; and anhydrous ammonia use by farmers. These operations typically involve a relative large number of small facilities that normally are not involved in chemical process safety. These facilities traditionally have not had the awareness, technical staff, or the level of technology necessary to deal with sophisticated PSM concepts (McKelvey et. al., 1992).

4.4 PROCESS HAZARD ANALYSIS: SPECIFIC REQUIREMENTS

Under the Clean Air Act Amendments, section 304(c)(2), OSHA must require employers to perform a workplace hazard assessment (OSHA's process hazard analysis), including, as appropriate, identification of potential sources of accidental release, an identification of any previous release within the facility which had a likely potential for catastrophic consequences in the workplace, estimation of workplace effects of a range of releases, and an estimation of the health and safety effects of such ranges on employees.

The final standard must meet the requirements contained in section 304(c) (2), (4), and (5) of the Clean Air Act Amendments. The requirements state that the OSHA standard must require employers to:

> (2) Perform a workplace hazard assessment (OSHA's Process Hazard Analysis) including, as appropriate, identification of potential sources of accidental release, an identification of any previous release within the facility

which had a likely potential for catastrophic consequences in the workplace, estimation of workplace effects of such range on employees.

(4) Establish a system to respond to the workplace hazard assessment findings, which shall address prevention, mitigation, and emergency responses.

(5) Periodically review the workplace hazard assessment and response system.

The process hazard analysis is a key component of a process safety management system because it is a thorough, orderly, systematic approach for identifying, evaluating and controlling processes involving highly hazardous chemicals.

In paragraph (e)(1) of the rule, OSHA requires that employers conduct an initial process hazard analysis on facilities covered by the standard in order to identify, evaluate and control the hazards of the process. By properly performing a hazard analysis, the employer can determine where problems may occur, take corrective measures to improve the safety of the process and pre-plan the actions that would be necessary if there were a failure of safety controls or other failures in the process. An employer needs to select a process hazard analysis method that is appropriate to the complexity of the process being analyzed.

OSHA requires that employers determine and document the priority order for conducting process hazard analyses based on such considerations as the extent of the process hazards, number of potentially affected employees, age of the process, and operating history of the process. This requirement is written flexibly in recognition of the fact that different processes will require different considerations for prioritization. The PSM Standard language indicates that process hazard analyses must be conducted as soon as possible.

In recognition that facilities will need time to compile the information required in paragraph (d), process safety information, which is needed to conduct a process hazard analysis, OSHA adopted a schedule that requires at least 25 percent of the process hazard analyses to be completed each year, starting with the second year after the effective date of the standard. OSHA chose to grandfather process hazard analyses completed five years before the effective date of the standard. These process hazard analyses must meet the requirements contained in paragraph (e) and will have to be updated and revalidated, based on their completion date, in accordance with the requirements in paragraph (e)(6). OSHA felt that it would not be reasonable to require that resources be expended to conduct another process hazard analysis when a recent one already exists since these same resources could be better used to conduct initial process hazard analyses on other processes.

The regulations also take a performance-oriented approach with respect to the process hazard analysis so that an employer has flexibility in choosing the type of analysis that would best address a particular process. In paragraph (e)(1) of the regulations, OSHA permits an employer to use one or more of certain listed methodologies to perform a process hazard analysis. The methodologies included: what-if; checklist; what-if/checklist; failure mode and effects analysis; hazard and operability study; and fault free analysis. OSHA felt that any methodology should be allowed as long as it meets the specified criteria described in paragraph (e).

Paragraph (e)(3) of the PSM Standard states what a hazard analysis must address. The analysis must address the hazards of the process; engineering and administrative controls applicable to the hazards and their interrelationships; the consequences of failure of these controls; and a consequence analysis of the effects of a release on all workplace employees. Paragraph (e)(2) concerning what a hazard analysis must address also requires identification of any previous incident which had a likely potential for catastrophic consequences. The inclusion of previous incidents is intended to help ensure that the process hazard analysis adequately addresses a wide enough range of concerns.

Paragraph (e)(3)(iii) of the rule requires that the process hazard analysis address: Engineering and administrative controls applicable to the hazards and their interrelationships, such as the appropriate application of detection methodologies to provide early warning of releases. (Acceptable detection methods might include process monitoring and control instrumentation with alarms, and detection hardware such as hydrocarbon sensors.) It should be noted, however, that OSHA used detection methodologies only as an example. There may be many other interrelationships that must be covered to comply with this provision for a particular process.

Chapter 4 - Process Hazard Analysis 91

In paragraph (e)(3)(iv), OSHA requires that the process hazard analysis address "consequence of failure of engineering and administrative controls." Paragraph (e)(3)(vii) requires a qualitative evaluation of the possible safety and health effects of failure of engineering and administrative controls on employees in the workplace. This evaluation is for the purpose of guiding decisions and priorities in planning for prevention and control, mitigation and emergency response.

OSHA felt that facility siting should always be considered during process hazard analyses. In order to ensure that employers consider siting, OSHA decided to specifically require it in paragraph (e)(3)(v). Finally, paragraph (e)(3)(vi) to the PSM rule requires that employers address human factors in the process hazard analysis.

4.5 TEAM APPROACH

Paragraph (e)(4) requires that employers conduct a process hazard analysis using a team approach. OSHA felt that in order to conduct an effective, comprehensive process hazard analysis, it is imperative that the analysis be performed by competent persons, knowledgeable in engineering and process operations, and those persons be familiar with the process being evaluated. Some employers may have a staff with expertise to perform a process hazard analysis. This staff will already be familiar with the process being evaluated. However, some companies, particularly smaller ones, may not have the staff expertise to perform such an analysis. The employer, therefore, may need to hire an engineering or consulting company to perform the analysis. It is important to note that in all situations, the team performing the process hazard analysis must include at least one employee from the facility who is intimately familiar with the process.

A team approach is the best approach for performing a process hazard analysis. This is because no one person will possess all of the knowledge and experience necessary to perform an effective process hazard analysis. Additionally, when more than one person is performing the analysis, different disciplines, opinions, and perspectives will be represented and additional knowledge and expertise will be contributed to the analysis. In fact, some companies even include an individual on the team who does not have any prior experience with the particular process being analyzed to help ensure that a fresh view of the process is integrated into the analysis. Employees and other experts may be brought onto the team on a temporary basis to contribute their specialized knowledge to the conduct of the process hazard analysis.

The PSM regulations require that the process hazard analysis be performed by a team with members who are knowledgeable in engineering and process operations, and that the team have at least one employee who has experience and knowledge specific to the process being evaluated.

A great number of rulemaking participants objected to the inclusion of an employee representative (union representative) on these teams, so OSHA decided not to specifically require an employee representative on the team. Instead, the agency chose to include a separate paragraph addressing employee participation in the overall process safety management program. This paragraph (c) requires employee participation in the process hazard analysis by requiring that employers consult with employees and their representatives on the conduct and development of process hazard analyses. Still, OSHA does require that an employee who has experience and knowledge specific to the process being evaluated be included on the team.

In paragraph (e)(5), the employer must assure that the recommendations resulting from the process hazard analysis are "resolved" in a timely manner and that the resolution is documented. Many times, on further study, process hazard analysis team recommendations are resolved in more effective ways than those originally envisioned by the team. OSHA felt that the employer should be given the option to implement solutions that are more effective than those recommended by the team. It is also possible that not all team recommendations will be correct or will resolve the problem found in the best way. When a team recommendation is incorrect, the employer has the opportunity to analyze it and then document in writing why the recommendation is not being adopted or is being adopted with modification.

Under paragraph (e)(6), the process hazard analysis must be updated and revalidated at least every five years, using the process hazard analysis team to ensure that the process hazard analysis is consistent with the current process. OSHA felt that this five-year update and revalidation interval was a reasonable timeframe, particularly in consideration of the long lifespan, without change, of many processes. Paragraph (e)(7) requires that employers retain the process hazard analysis and their updates and revalidation. OSHA felt that this requirement would not pose an undue burden on employers since retention of these documents is necessary to conduct the periodic updates and revalidations required under the standard.

4.6 PROCESS HAZARD ANALYSIS: COMPLIANCE GUIDELINES

A process hazard analysis (PHA), sometimes called a process hazard evaluation, is one of the most important elements of the process safety management program. A PHA is an organized and systematic effort to identify and analyze the significance of potential hazards associated with the processing or handling of highly hazardous chemicals. A PHA provides information which will assist employers and employees in making decisions for improving safety and reducing the consequences of unwanted or unplanned releases of hazardous chemicals. A PHA is directed toward analyzing potential causes and consequences of fires, explosions, releases of toxic or flammable chemicals and major spills of hazardous chemicals. The PHA focuses on equipment, instrumentation, utilities, human actions (routine and nonroutine), and external factors that might impact the process. These considerations assist in determining the hazards and potential failure points or failure modes in a process.

The selection of a PHA methodology or technique will be influenced by many factors including the amount of existing knowledge about the process. Is it a process that has been operated for a long period of time with little or no innovation and extensive experience has been generated with its use? Or, is it a new process or one which has been changed frequently by the inclusion of innovative features? Also, the size and complexity of the process will influence the decision as to the appropriate PHA methodology to use. All PHA methodologies are subject to certain limitations. For example, the checklist methodology works well when the process is very stable and no changes are made, but it is not as effective when the process has undergone extensive change. The checklist may miss the most recent changes and consequently the changes would not be evaluated. Another limitation to be considered concerns the assumptions made by the team or analyst. The PHA is dependent on good judgement and the assumptions made during the study need to be documented and understood by the team and reviewer and kept for a future PHA.

94   OSHA and EPA PSM Requirements

4.6.1   Process Hazard Analysis Compliance Checklist

1910.119 (e): PROCESS HAZARD ANALYSIS

I. PROGRAM SUMMARY

The intent of this paragraph is to require the employer to develop a thorough, orderly, systematic approach for identifying, evaluating and controlling processes involving highly hazardous chemicals. Minimum requirements include: (1) Setting a priority order and conducting analyses according to the required schedule; (2) Using an appropriate methodology to determine and evaluate the process hazards; (3) Addressing process hazards, previous incidents with catastrophic potential, engineering and administrative controls applicable to the hazards, consequences of failure of controls, facility siting, human factors, and a qualitative evaluation of possible safety and health effects of failure of controls on employees; (4) Performing PHA by a team with expertise in engineering and process operations, the process being evaluated, and the PHA methodology used; (5) Establishing a system to promptly address findings and recommendations, assure recommendations are resolved and documented, document action taken, develop a written schedule for completing actions, and communicate actions to operating, maintenance and other employees who work in the process or might be affected by actions; (6) Updating and revalidating PHA's at least every 5 years; and (7) Retaining PHA's and updates for the life of the process.

II. QUALITY CRITERIA REFERENCES

A.   1910.119(e):   Process Hazard Analysis

III. VERIFICATION OF PROGRAM ELEMENTS | Criteria Reference | Met Y/N

A.   Records Review

1. Has the employer determined and documented a priority order for conducting initial PHA's based on a rationale that includes at least these factors:
 • the extent of process hazards
 • number of potentially affected employees
 • age of process
 • operating history?

.119(e)
(1)

FIELD NOTE REFERENCE(S):

Chapter 4 - Process Hazard Analysis

2. Are the initial PHA's for processes covered by the PSM standard being performed as soon as possible? FIELD NOTE REFERENCE(S):	.119(e)(1)	
3. Does the priority schedule for PHA's assure that all initial PHA's will be performed by 5/26/97 and that: • No less than 25% of the PHA's shall be completed by 5/26/94? • No less than 50% of the PHA's shall be completed by 5/26/95? • No less than 75% of the PHA's shall be completed by 5/26/96? (PHA's completed after May 26, 1987 which meet the requirements of this paragraph are acceptable as initial PHA's; they must be updated and revalidated at least every 5 years.) FIELD NOTE REFERENCE(S):	.119(e)(1)	
4. Does the hazard evaluation use one or more of the following PHA methodologies: • What-if? • Checklist? • What-if/Checklist? • Hazard & Operability Study (HAZOP)? • Failure Mode and Effects Analysis (FMEA)? • Fault Tree Analysis (FTA)? • Other appropriate methodology? (See Appendix B for a discussion of appropriate methodologies). FIELD NOTE REFERENCE(S):	.119(e)(2)	

5. Does the PHA address the following: • The hazards of the process? FIELD NOTE REFERENCE(S): • Previous incidents with likely potential for catastrophic consequences? FIELD NOTE REFERENCE(S): • Consequences of failure of engineering and administrative controls? (For example, potential injury, maximum release of hazardous materials, property damage, etc.) FIELD NOTE REFERENCE(S):	.119(e)(3)

Chapter 4 - Process Hazard Analysis   97

5. (Continued) Does the PHA address the following: • Engineering and administrative controls applicable to the hazards and their interrelationships? (Such controls may include appropriate application of detection methodologies to provide early warning of releases; inventory reduction; substitution of less hazardous materials; protective systems such as deluges, monitors, foams; increased separation distances; modification of the process temperature or pressure; redundancy in instrumentation; etc.) FIELD NOTE REFERENCE(S): • Facility siting? (Review calculations, charts, and other documents that verify facility siting has been considered. For example, safe distances for locating control rooms may be based on studies of the individual characteristics of equipment involved such as: types of construction of the room, types and quantities of materials, types of reactions and processes, operating pressures and temperatures, presence of ignition sources, fire protection facilities, capabilities to respond to explosions, drainage facilities, location of fresh air intakes, etc.) FIELD NOTE REFERENCE(S): • Human factors? (Such factors may include a review of operator/process and operator/equipment interface, the number of tasks operators must perform and the frequency, the evaluation of extended or unusual work schedules, the clarity and simplicity of control displays, automatic instrumentation versus manual procedures, operator feedback, clarity of signs and codes, etc.) FIELD NOTE REFERENCE(S): • A qualitative evaluation of a range of possible safety and health effects of failure of controls on employees in the workplace? FIELD NOTE REFERENCE(S):	.119(e)(3)	

98   OSHA and EPA PSM Requirements

6. Are the process hazard analyses performed by teams with expertise in engineering and process operations, including at least one employee with experience and knowledge specific to the process being evaluated and one member knowledgeable in the specific PHA methodology used? FIELD NOTE REFERENCE(S):	.119(e)(4)	
7. Has a system been established to promptly address the team's findings and recommendations? FIELD NOTE REFERENCE(S): Review a representative sample of the documentation. Has the system been able to: • Assure that the recommendations are resolved and documented in a timely manner? FIELD NOTE REFERENCE(S): • Document actions to be taken? FIELD NOTE REFERENCE(S): • Complete actions as soon as possible? FIELD NOTE REFERENCE(S): • Develop a written schedule of when actions are to be completed? FIELD NOTE REFERENCE(S): • Communicate the actions to operating, maintenance and other employees whose work assignments are in the process and who may be affected by the recommendations or actions? FIELD NOTE REFERENCE(S):	.119(e)(5)	
8. Are the PHA's updated and revalidated at least every five years by a qualified team meeting the requirements in paragraph (e)(4), to assure that the process hazard analysis is consistent with the current process? FIELD NOTE REFERENCE(S):	.119(e)(6)	

9. Are all initial PHA's, updates or revalidations, and documented resolutions of recommendations kept for the life of the process? FIELD NOTE REFERENCE(S):	.119(e) (7)	
B. On-site Conditions 1. Do observations of a representative sample of process-related equipment indicate that obvious hazards have been identified, evaluated, and controlled? (For example, hydrocarbon or toxic gas monitors and alarms are present; electrical classifications are consistent with flammability hazards; destruct systems such as flares are in place and operating; control room siting is adequate or provisions have been made for blast resistant construction, pressurization, alarms, etc.; pressure relief valves and rupture disks are properly designed and discharge to a safe area; pipework is protected from impact; etc.) FIELD NOTE REFERENCE(S):	.119(e) (1)	
2. Do observations of a representative sample of process-related equipment indicate that PHA recommendations have been promptly resolved? FIELD NOTE REFERENCE(S):	.119(e) (5)	
C. Interviews PHA Team Members: 1. Based on interviews with a representative number of the PHA team members, are the PHA methodologies used appropriate for the complexity of the process? FIELD NOTE REFERENCE(S):	.119(e) (1)	
2. Based on interviews with a representative number of the PHA team members, is the priority order for conducting PHA's based on the extent of the process, the number of potentially affected employees, the age of the process, and the operating history of the process? FIELD NOTE REFERENCE(S):	.119(e) (1)	

100 OSHA and EPA PSM Requirements

3. Based on interviews with a representative number of the PHA team members, have the following been addressed: • The hazards of the process? • Previous incidents with likely potential for catastrophic consequences? • Engineering and administrative controls applicable to the hazards? • Consequences of control failures? • Facility siting? • Human factors? (Ask about shift rotations, extended schedules, and other possible sources of error.) • A qualitative evaluation of a range of possible safety and health effects of failure of controls on employees in the workplace? FIELD NOTE REFERENCE(S):	.119(e)(3)
4. Based on interviews with a representative number of the PHA team members, do the members have the appropriate expertise in engineering, process operations, and the process methodology used? Does one member of the team have experience and knowledge in the specific process? FIELD NOTE REFERENCE(S):	.119(e)(4)
5. Based on interviews with a representative number of the PHA team members, does the system established by the employer address the team's findings and recommendations promptly? FIELD NOTE REFERENCE(S):	.119(e)(5)
Operators and maintenance: 6. Based on interviews with a representative number operator and maintenance employees, have the PHA's addressed the recognized hazards of the process and previous incidents which had a likely potential for catastrophic consequences? FIELD NOTE REFERENCE(S):	.119(e)(3)
7. Based on interviews with operator, maintenance, and other employees who may be affected by PHA recommendations, have actions taken to resolve PHA findings been communicated to these employees? FIELD NOTE REFERENCE(S):	.119(e)(5)

4.6.2 PHA Team

The team conducting the PHA needs to understand the methodology that is going to be used. A PHA team can vary in size from two people to a number of people with varied operational and technical backgrounds. Some team members may only be a part of the team for a limited time. The team leader needs to be fully knowledgeable in the proper implementation of the PHA methodology that is to be used and should be impartial in the evaluation. The other full- or part-time team members need to provide the team with expertise in areas such as process technology, process design, operating procedures and practices, including how the work is actually performed, alarms, emergency procedures, instrumentation, maintenance procedures, both routine and nonroutine tasks, including how the tasks are authorized, procurement of parts and supplies, safety and health, and any other relevant subject as the need dictates. At least one team member must be familiar with the process.

The ideal team will have an intimate knowledge of the standards, codes, specifications and regulations applicable to the process being studied. The selected team members need to be compatible and the team leader needs to be able to manage the team, and the PHA study. The team needs to be able to work together while benefiting from the expertise of others on the team or outside the team, to resolve issues, and to forge a consensus on the findings of the study and recommendations.

4.6.3 Application of PHA to Specific Processes

The application of a PHA to a process may involve the use of different methodologies for various parts of the process. For example, a process involving a series of unit operations of varying sizes, complexities, and ages may use different methodologies and team members for each operation. Then the conclusions can be integrated into one final study and evaluation. A more specific example is the use of a checklist PHA for a standard boiler or heat exchanger and the use of a Hazard and Operability PHA for the overall process. Also, for batch type processes like custom batch operations, a generic PHA of a representative batch may be used where there are only small changes of monomer or other ingredient ratios and the chemistry is documented for the full range and ratio of batch ingredients. Another process that might

102 OSHA and EPA PSM Requirements

consider using a generic type of PHA is a gas plant. Often these plants are simply moved from site to site and therefore, a generic PHA may be used for these movable plants. Also, when an employer has several similar size gas plants and no sour gas is being processed at the site, then a generic PHA is feasible as long as the variations of the individual sites are accounted for in the PHA. Finally, when an employer has a large continuous process which has several control rooms for different portions of the process such as for a distillation tower and a blending operation, the employer may wish to do each segment separately and then integrate the final results.

4.6.4 Small Businesses

Additionally, small businesses which are covered by this rule will often have processes that have less storage volume, less capacity, and are less complicated than processes at a large facility. Therefore, OSHA would anticipate that the less complex methodologies would be used to meet the process hazard analysis criteria in the standard. These process hazard analyses can be done in less time and with only a few people being involved. A less complex process generally means that less data, P&IDs, and process information is needed to perform a process hazard analysis.

Many small businesses have processes that are not unique, such as cold storage lockers or water treatment facilities. Where employer associations have a number of members with such facilities, a generic PHA, evolved from a checklist or what-if questions, could be developed and used by each employer effectively to reflect his/her particular process; this would simplify compliance for them. When the employer has a number of processes which require a PHA, the employer must set up a priority system of which PHAs to conduct first. A preliminary or gross hazard analysis may be useful in prioritizing the processes that the employer has determined are subject to coverage by the process safety management standard. Consideration should first be given to those processes with the potential of adversely affecting the largest number of employees. This prioritizing should consider the potential severity of a chemical release, the number of potentially affected employees, the operating history of the process such as the frequency of chemical releases, the age of the process and any other relevant factors. These factors would suggest a ranking order and would suggest either using a weighing factor system or a systematic ranking method. The use of a preliminary hazard analysis would assist an employer in determining which process should be of the highest priority and thereby the employer would obtain the greatest improvement in safety at the facility. More detailed guidance on the content and application of process hazard analysis methodologies is available from the American Institute of Chemical Engineers' Center for Chemical Process Safety.

Chapter 5

Training Employees and Using Contractors

5.1 EMPLOYEE TRAINING PROGRAMS

To keep complex process equipment and machinery operating safely, personnel at all levels in the organization need proper training. At the most basic levels, operators and mechanics need training to understand and implement operating rules and procedures. Personnel also need training to understand and properly conduct process hazard analyses. Managers need training to learn the most efficient methods for overseeing and managing the entire process. Without proper training, the chances of safe process operation are seriously diminished, even if all other key process safety management elements are in place.

The PSM Standard requires that employers train each employee involved in operating a process with an overview of the process and its operating procedures. The training must include emphasis on the specific safety and health hazards of the process, emergency operations, and other safe work practices that apply to the employee's job tasks. Those employees involved in operating a process on the PSM Standard's effective date do not necessarily need to be given an initial training. The training requirement may be satisfied by a written certification by the employer stating that these employees have the required knowledge, skills and abilities to carry out the process safely in accordance with the operating procedures. Still, refresher training must be provided at least every three years to ensure that each employee involved with operating a process understands and adheres to the current operating procedures for the process. The employer also must maintain records documenting that each employee operating a process has received and understood the PSM training. Records must identify each employee, state when training was given, and show how the employer verified that the employee understood the training.

5.2 TRAINING: SPECIFIC REQUIREMENTS

The implementation of an effective training program is one of the most important steps that an employer can take to enhance employee safety. An effective training program will help employees understand the nature and causes of problems arising from process operations, and will increase employee awareness with respect to the hazards particular to a process. An effective training program will significantly reduce the number and severity of incidents arising from process operations, and can be instrumental in preventing small problems from leading to a catastrophic release.

Paragraph (g)(1) covers initial training, and required each employee presently "involved" in a process, and each new employee before working in a newly assigned process, to be trained in an overview of the process and in the operating procedures specified in paragraph (f) of the regulations. This provision proposal also requires the training to include emphasis on the specific safety and health hazards, procedures, and safe work practices applicable to the employee's job tasks.

Paragraph (g)(1) covers three broad topics related to initial training: the application of this provision (to whom the training applies); OSHA's approach (including the amount and method of training, and the content of the training program); and grandfathering of training (the recognition of training received by employees prior to promulgation of this standard).

This provison reads as follows:
"Each employee presently involved in operating a process, and each employee before being involved in operating a newly assigned process, shall be trained in an overview of the process and in the operating procedures as specified in paragraph (f) of this section. The training shall include emphasis on the specific safety and health hazards, emergency operations including shutdown, and safe work practices applicable to the employee's job tasks."

5.2.1 Application

Paragraph (g)(1) applies to "employees involved in operating a process." OSHA meant for this provision to apply to only those employees, including managers and supervisors, who are actually involved in "operating" the process. While most OSHA standards, by their terms, apply to all employees in a particular situation and contract employees are considered "employees" in the broad sense of the word, this standard distinguishes in the training requirements between contract employees and direct hire employees. This was done primarily for emphasis and in recognition of the fact that in some segments of industry covered by the process safety management standard, contractors make up a substantial portion of on-site workers. OSHA wanted to focus attention on that situation and did so by imposing separate but similar training objectives for direct hire and contract employees. For this reason, training requirements for contractor employees and maintenance employees are addressed separately in paragraph (h) of the PSM rule.

After a careful analysis, OSHA concluded that a performance-oriented approach to training is appropriate. OSHA felt that employers can determine the amount of training and the content of the training program that best reflects the operation's complexity and the experience and necessary skill level of their employees.

5.2.2 Grandfathering

Many rulemaking participants contended that OSHA should recognize training that employees received prior to the promulgation of this standard. Therefore, OSHA decided to allow grandfathering of initial training under certain circumstances. Thus, paragraph (g)(1)(ii) reads as follows:
"In lieu of initial training for those employees already involved in operating a process on May 26, 1992, an employer may certify in writing that the employee has the required knowledge, skills, and abilities to safely carry out the duties and responsibilities as specified in the operating procedures."

5.2.3 Refresher Training

Paragraph (g)(2) requires refresher training be provided to each employee at least every three years, and more often as necessary, to ensure that the employee understands and adheres to the current operating procedures of the process. Accordingly, paragraph (g)(2) in the PSM rule reads as follows:
"Refresher training shall be provided at least every three years, and more often if necessary, to each employee involved in operating a process to assure that the employee understands and adheres to the current operating procedures of the process. The employer, in consultation with the employees involved in operating the process, shall determine the appropriate frequency of refresher training."

OSHA determined that the employer, in consultation with employees, can best determine the appropriate frequency of refresher training. While OSHA also concluded that it would be inappropriate to prescribe a minimum number of hours of refresher training because there is a wide variation in operation complexity, and in the experience and skill levels of employees, OSHA felt that the frequency of refresher training should be held at least every three years to assure that employees understand and adhere to current operating procedures.

5.2.4 Training Documentation

Paragraph (g)(3) requires the employer to document that employees have received and successfully completed the required training. It also requires the documentation to identify the employee, the date of the training, and the signature of the person doing the training. Paragraph (g)(3), "Training documentation," reads as follows:
"The employer shall document that each employee involved in operating a process has received and understood the training required by this paragraph. The employer shall prepare a record which contains the identity of the employee, the date of training, and the means used to verify that the employee understood the training."

The purpose of this provision is to ensure that employees not only receive training but, also, that they understand and can demonstrate what they have learned in order to perform their job tasks safely. This is especially important when comprehensive training and the understanding of the training plays such a crucial role in the risk reduction associated with the process safety management rule. OSHA also felt that this provision was necessary to serve as a tracking mechanism for the training that employees receive and when employees received the training.

Consistent with the performance-oriented approach of the PSM Standard, any one of several methods, or combination of methods, can be used to verify that employees understand the training that they have received. Employers are therefore free to devise the method that works best in their establishment to ascertain that employees have understood their training. Consequently, OSHA did not mandate any specific methods of training validation in the regulations.

Additionally, OSHA felt that it was important that the training documentation contain the name of the person conducting the training, as opposed to the signature of the person conducting the training as was proposed. This allows employers to keep training records on computer if they so desire.

5.2.5 Clean Air Act Training Requirements

Section 304(c)(9) of the Clean Air Act Amendments mandated that this standard contain a provision requiring employers to "train and educate employees and contractors in emergency response in a manner as comprehensive and effective as that required by the regulation promulgated pursuant to section 126(d) of the Superfund Amendments and Reauthorization Act" (SARA). That section of SARA requires that workers receive a specified minimum number of hours of training unless the worker "has received the equivalent of such training."

It is OSHA's position that the training requirements contained in paragraph (g) of the PSM rule, together with the requirements pertaining to emergency planning and response contained in paragraph (n) of the rule provide "equivalent training" to the training required for emergency response under section 126(d) of SARA. In addition, those employees who would be involved in emergency response must meet the training requirements in 29 CFR 1910.120, Hazardous Waste Operations and Emergency Response, referenced in paragraph (n) of the PSM rule, which is directly responsive to section 126(d) of SARA.

5.3 EMPLOYEE TRAINING COMPLIANCE GUIDELINES

All employees, including maintenance and contractor employees, involved with highly hazardous chemicals need to fully understand the safety and health hazards of the chemicals and processes they work with for the protection of themselves, their fellow employees and the citizens of nearby communities. Training conducted in compliance with 29 CFR 1910.1200, the Hazard Communication Standard, will help employees to be more knowledgeable about the chemicals they work with, as well as familiarize them with reading and understanding MSDSs. However, additional training in subjects such as operating procedures and safety work practices, emergency evacuation and response, safety procedures, routine and nonroutine work authorization activities, and other areas pertinent to process safety and health will need to be covered by an employer's training program.

In establishing their training programs, employers must clearly define the employees to be trained and what subjects are to be covered in their training. Employers in setting up their training program will need to clearly establish the goals and objectives they wish to achieve with the training that they provide to their employees. The learning goals or objectives should be written in clear measurable terms before the training begins. These goals and objectives need to be tailored to each of the specific training modules or segments. Employers should describe the important actions and conditions under which the employee will demonstrate competence or knowledge as well as what is acceptable performance.

5.3.1 Training Examples

Hands-on training, where employees are able to use their senses beyond listening, will enhance learning. For example, operating personnel, who will work in a control room or at control panels, would benefit by being trained at a simulated control panel or panels. Upset conditions of various types could be displayed on the simulator, and then the employee could go through the proper operating procedures to bring the simulator panel back to the normal operating parameters. A training environment could be created to help the trainee feel the full reality of the situation but, of course, under controlled conditions. This realistic type of training can be very effective in teaching employees correct procedures while allowing them to also see the consequences of what might happen if they do not follow established operating procedures. Other training techniques using videos or on-the-job training can also be very effective for teaching other job tasks, duties, or other important information. An effective training program will allow the employee to fully participate in the training process and to practice their skill or knowledge.

5.3.2 Training Program Evaluation

Employers need to periodically evaluate their training programs to see if the necessary skills, knowledge, and routines are being properly understood and implemented by their trained employees. The means or methods for evaluating the training should be developed along with the training program goals and objectives. Training program evaluation will help employers to determine the amount of training their employees understood, and whether the desired results were obtained. If, after the evaluation, it appears that the trained employees are not at the level of knowledge and skill that was expected, the employer will need to revise the training program, provide retraining, or provide more frequent refresher training sessions until the deficiency is resolved. Those who conducted the training and those who received the training should also be consulted as to how best to improve the training process. If there is a language barrier, the language known to the trainees should be used to reinforce the training messages and information.

Careful consideration must be given to ensure that employees including maintenance and contract employees receive current and updated training. For example, if changes are made to a process, impacted employees must be trained in the changes and understand the effects of the changes on their job tasks (e.g., any new operating procedures pertinent to their tasks).

5.3.3 Training Compliance Checklist

1910.119 (g): TRAINING

I. PROGRAM SUMMARY

The intent of this paragraph helps employees understand the nature and causes of problems arising from process operations, and increases employee awareness with respect to the hazards particular to a process. An effective training program significantly reduces the number and severity of incidents arising from process operations, and can be instrumental in preventing small problems from leading to a catastrophic release. Minimum requirements for an effective training program include: Initial Training, Refresher Training, and Documentation.

II. QUALITY CRITERIA REFERENCES

    A.   1910.119(g): Training
    B.   1910.119(f)(1): Operating procedures

III. VERIFICATION OF PROGRAM ELEMENTS	Criteria Reference	Met Y/N
A. Records Review 1. For employees involved in operating a process do initial and refresher training records exist? Do the records contain the identity of the employee, the date of the training, and the means used to verify that the employee understood the training? FIELD NOTE REFERENCE(S):	.119(g)(1)(i)	
2. Has each employee been trained before being involved in a newly assigned process (except employees involved in operating a process prior to 5/26/92)? FIELD NOTE REFERENCE(S):	.119(g)(1)(i)	

3. If initial training has not been given to employees involved in operating a process prior to 5/26/92, is there written certification that they have the required knowledge, skills and abilities to safely carry out the duties and responsibilities specified in the operating procedures? (Review the documents to make sure the certification has not be invalidated by a change in duties.) FIELD NOTE REFERENCE(S):	.119(g)(1)(ii)	

4. Has each employee involved in operating a process been trained in an overview of the process and the operating procedures including:
• Steps for each operating phase?
    Initial startup, normal operations, temporary operations, emergency shutdown, emergency operations, normal shutdown, and startup following a turnaround or emergency shutdown
• Operating limits?
    Consequences of deviations and steps required to avoid deviations
• Safety and health considerations?
    Properties and hazards of chemicals used and precautions for preventing exposure
• Safety systems and their functions?
 	
FIELD NOTE REFERENCE(S):
	.119(g)(1)(i)	

5. Has the employer consulted with employees involved in operating the process to determine the appropriate frequency for refresher training? Is the frequency at least once every 3 years? FIELD NOTE REFERENCE(S):	.119(g)(2)	
B. On-site Conditions Verification is not required.	.119(g)(1) or (2)	

C. Interviews

1. Based on interviews with a representative number of employees, has their training emphasized specific safety and health hazards, emergency operations including shutdown, and safe work practices applicable to their tasks? FIELD NOTE REFERENCE(S):	.119(g)(1)(i)	
2. Based on interviews with employees named as having provided consultation, has the employer consulted with employees involved in operating the process to determine the appropriate frequency of refresher training? FIELD NOTE REFERENCE(S):	.119(g)(2)	

114 OSHA and EPA PSM Requirements

5.4 APPLICATION OF PSM STANDARD TO CONTRACTORS

Shortly after the catastrophic Phillips 66 Company's Houston Chemical Complex incident in 1989, OSHA asked the John Gray Institute (JGI) of Lamar University to conduct a study of safety and health issues as they relate to contract work in the petrochemical industry. The issue of the role of contractors in the petrochemical industry surfaced since a contractor had been working in the vicinity of the Phillips' release. Additionally, OSHA determined that a significant number of companies were using contractors to perform work at their plants. OSHA wanted to ensure that safety issues surrounding contractor employees who are exposed or may expose site employees to potentially catastrophic events were thoroughly addressed in the process safety management standard. Upon the completion of the JGI report, OSHA decided to give interested persons an opportunity to comment on the report and, therefore, reopened the rulemaking record on the PSM Standard to receive public comment on the report and to reexamine the provisions concerning contractors. (See 56 Fed. Reg. 48133 (Sept. 24, 1991).) OSHA received more than 300 requests for the JGI report.

On September 24, 1991, OSHA published a notice in the Federal Register announcing the availability of the JGI report on contractors and peer review of the report. The public was given an opportunity to comment and reexamine the contractor provisions of the proposed process safety management standard in light of the JGI report (56 FR 48133). The comment period ended on October 24, 1991, and OSHA received 37 comments in response to the notice. Generally commenters viewed some of the issues addressed in the report as important considerations. However, many commenters expressed their belief that the report should not be used as a basis in the development of the final contractor provisions in the final process safety management standard. Commenters questioned the credibility of the report's findings and recommendations and pointed to criticisms leveled at the report by its peer reviewers and the criticisms that resulted from the special evaluation of the final JGI report conducted for The Business Roundtable by the University of Texas at Austin and Texas A&M University. The evaluation concluded that the JGI report's "conclusions are based on a highly problematic research design, research methodologies, data, analysis of data, and intepretation of results" and further observed that the review teams (one from the University of Texas at Austin and one from Texas A&M University) "are unanimous in concluding that the JGI report should be treated with extreme caution and should not be used as a basis for establishing national policy or industry standards." Additionally, some commenters observed that the JGI report only dealt with the petrochemical industry and that OSHA should not use it to draw conclusions with regard to other industry segments covered by the process safety management standard.

OSHA did not use the final JGI report as a basis for requirements in the development of its final provisions concerning contractors. A review of the comments in the record indicated that significant other information and data is available on which the final contractor provisions can be based. While OSHA decided not to use the report as a basis for the final contractor provisions, OSHA felt that the final provisions of the PSM rule benefited by the additional public input which reconfirmed, clarified and expanded on comments and testimony received.

5.4.1 Use of Contractors

Many categories of contract labor may be present at a jobsite; such workers may actually operate the facility or do only a particular aspect of a job because they have specialized knowledge or skill. Others work only for short periods when there is an immediate need for increased staff, such as in turnaround operations. The PSM rule includes special provisions for contractors and their employees to emphasize the importance of everyone taking care that they do nothing to endanger those working nearby who may work for another employer. The PSM Standard, therefore, applies to contractors performing maintenance or repair, turnaround, major renovation, or specialty work on or adjacent to a covered process. It does not apply, however, to contractors providing incidental services that do not influence process safety, such as janitorial, food and drink, laundry, delivery, or other supply services.

5.4.2 Employer Responsibilities

When selecting a contractor, the employer must obtain and evaluate information regarding the contract employer's safety performance and programs. The employer also must inform contract employers of the known potential fire, explosion, or toxic release hazards related to the contractor's work and the process; explain to the contract employers the applicable provisions of the emergency action plan; develop and implement safe work practices to control the entrance, presence, and exit of contract employers and contract employees in covered process areas; evaluate periodically the performance of contract employers in fulfilling their obligations; and maintain a contract employee injury and illness log related to the contractor's work in process areas.

5.4.3 Contract Employer Responsibilities

The contract employer also has to fulfill several requirements under the PSM Standard. The contract employer must:

* Ensure that contract employees are trained in the work practices necessary to perform their job safely;
* Ensure that contract employees are instructed in the known potential fire, explosion, or toxic release hazards related to their job and the process, and in the applicable provisions of the emergency action plan;
* Document that each contract employee has received and understood the training required by the PSM Standard by preparing a record that contains the identity of the contract employee, the date of training, and the means used to verify that the employee understood the training;
* Ensure that each contract employee follows the safety rules of the facility including the required safe work practices required in the operating procedures section of the PSM Standard; and
* Advise the employer of any unique hazards presented by the contract employer's work.

5.5 CONTRACTORS: SPECIFIC PSM REQUIREMENTS

Section 304 requirements of the Clean Air Act Amendments (CAAA) state that the OSHA standard must require employers to:

(8) Ensure contractors and contract employees are provided appropriate information and training.

Chapter 5 - Training / Contractors   117

(9) Train and educate employees and contractors in emergency response in a manner as comprehensive and effective as that required by the regulation promulgated pursuant to section 126(d) of the Superfund Amendments and Reauthorization Act.

OSHA believes that the contractor provisions contained in the final standard meet the requirements contained in section 304(c) (8) and (9) of the CAAA in a manner as comprehensive and effective as that required by the regulation promulgated pursuant to section 126(d) of the Superfund Amendments and Reauthorization Act for the reasons described in the preamble discussion regarding section 126(d) of the Superfund Amendments in paragraph (g), Training.

Paragraph (h) of the PSM Standard attempts to distinguish between the many types of contract workers who may be present at a jobsite and indicates the type of contract worker that the special training provisions of the regulation are attempting to cover. Among the many categories of contract labor that may be present at a particular jobsite, it is important to appreciate the differences among them. For example, contractors may actually operate a facility for an owner (who may own the facility but have little to do with the daily operation). In this case the contractor is the employer responsible for the covered processes and would obviously be treated as the "employer." Some contractors are hired to do a particular aspect of a job because they have a specialized area of expertise of which the host employer has little knowledge or skill (for example, asbestos removal). Other contractors work on site when the operation has need for increased manpower quickly for a short period of time, such as those involved in a turnaround operation. While paragraph (h)(2) sets forth the duties of the host employer to contract employers, the extent and the depth of these duties will depend to some degree on the category of contractor present. For example, should a contract employer provide employees to operate a process, then those employees would obviously have to be trained to the same extent as the directed hire employees "involved in operating a process" under paragraph (g) of the PSM Standard.

Generally speaking, all OSHA standards cover all employees including contract employees. In something of a break with tradition, the process safety management rule has separate

118 OSHA and EPA PSM Requirements

provisions covering the training of contract employees. This was done primarily for emphasis since contract employees make up a significant portion of some segment of industries covered by the PSM rule. This is not to say, however, that paragraph (h) is the only section of the process safety rule that applies to contractors. Under appropriate circumstances, all of the provisions of the standard may apply to a contractor (i.e., a contractor operated facility). After all, employees of an independent contractor are still employees in the broadest sense of the word and they and their employers must not only follow the process safety management rule, but they must also take care that they do nothing to endanger the safety of those working nearby who work for another employer. Moreover, the fact that this rule has a separate section that specifically lays out the duty of contractors on the jobsite does not mean that other OSHA standards, lacking a similar section, do not apply to contract employers.

OSHA has a long history of enforcing OSHA standards on multi-employer worksites. Nothing in this rule changes the position that the Agency has long taken in cases such as Anning-Johnson (4 O.S.H. Cas. (BNA) 1193), Harvey Workover, Inc. (7 O.S.H. Cas. (BNA) 1687) and in its Field Operations Manual (CPL 2.45B CH-1, Chapter V-9). As a general matter each employer is responsible for the health and safety of his/her own employees. However, under certain circumstances an employer may be cited for endangering the safety of another's employees. In determining who to hold responsible, OSHA will look at who created the hazard, who controlled the hazard and whether all reasonable means were taken to deal with the hazard.

Paragraph (h)(1) contains an application statement to clarify which contractors are covered by the standard. Thus, paragraph (h) covers those contractors whose work brings them into direct contact with, or whose work could affect the hazards of processes covered by the standard. OSHA decided that contractors providing incidental services are adequately covered under the 29 CFR 1910.1200, Hazard Communication Standard. Paragraph (h)(1) reads as follows:

> "(h) **Contractors.** (1) Application. This paragraph applies to contractors performing operating duties, maintenance or repair, turnaround, major renovation, or specialty work on or adjacent to a covered process area. It does not apply to contractors providing incidental services which do not

influence process safety, such as janitorial work, food and drink services, laundry, delivery or other supply services."

OSHA has delineated the responsibilities of employers and contractors in paragraph (h)(2), employer responsibilities, and paragraph (h)(3), contract employer responsibilities. The provisions concerning employer responsibilities read as follows:

"(2) Employer responsibilities. (i) The employer, when selecting a contractor, shall obtain and evaluate information regarding the contract employer's safety performance and programs.
(ii) The employer shall inform contract employers of the known potential fire, explosion, or toxic release hazards related to the contractors work and the process.
(iii) The employer shall explain to contract employers the applicable provisions of the emergency action plan required by paragraph (n) of this section.
(iv) The employer shall develop and implement safe work practices consistent with paragraph (f)(4) of this section, to control the entrance, presence and exit of contract employers and contract employees in process areas covered by this section.
(v) The employer shall periodically evaluate the performance of contract employers in fulfilling their obligations as specified in paragraph (h)(3).
(vi) The employer shall maintain a contract employee injury and illness log related to the contractor's work in process areas."

Paragraph (h)(2)(i) of the PSM Standard, requires that an employer, when selecting a contractor, obtain and evaluate information regarding a contractor employer's safety performance and programs. OSHA felt that an employer should be fully informed about a contract employer's safety performance. Therefore, the agency requires an evaluation of a contract employer's safety performance (e.g., an employer's experience modification rate) and safety programs. Evaluating safety performance and programs is an important measure in preserving the integrity of processes involving highly hazardous chemicals. OSHA anticipates that the requirement will provide employers an opportunity to ensure that they are not introducing additional

120 OSHA and EPA PSM Requirements

hazards to their processes; and will give employers an opportunity to request that contract employers improve their safety performance or make other adjustments to their safety programs in order to enhance the safety of all employees working in processes involving highly hazardous chemicals. The PSM rule, being performance oriented, does not require that employers refrain from using contractors with less than perfect safety records. However, the employer does have the duty to evaluate the contract employer's safety record and safety programs. Where the evaluation indicates some gaps in the contract employer's approach to safety, the employer may need to be more vigilant in the oversight and may need to develop and implement more stringent safe work practices to control the presence of contractors in covered process areas.

Paragraphs (h)(2) (ii) and (iii) of the PSM Standard require the communication of basic process hazard and emergency information to contract employers. Paragraph (h)(2)(iv) of the PSM Standard requires development and implementation of safe work practices which must be consistent with the operating procedures discussed in paragraph (f)(4) of the rule. Paragraph (h)(2)(v) of the PSM Standard requires employers to periodically evaluate the performance of contract employers in fulfilling their obligations. Finally, paragraph (h)(2)(vi) to the rule requires a log of injuries and illnesses to be kept by the employer. OSHA requires that an employer should be informed of all of the injuries and illnesses occurring in processes involving highly hazardous chemicals at the plant regardless of whether they be the employer's employees or the contractor's employees.

Paragraph (h)(3) of the PSM rule delineates the contract employer responsibilities and it includes the following provisions:

> "(3) Contract employer responsibilities. (i) The contract employer shall assure that each contract employee is trained in the work practices necessary to safely perform his/her job.
> (ii) The contract employer shall assure that each contract employee is instructed in the known potential fire, explosion, or toxic release hazards related to his/her job and the process, and the applicable provisions of the emergency action plan.
> (iii) The contract employer shall document that each contract employee has received and understood the training

required by this paragraph. The contract employer shall prepare a record which contains the identity of the contract employee, the date of training, and the means used to verify that the employee understood the training.

(iv) The contract employer shall assure that each contract employee follows all applicable work practices and safety rules of the facility including the safe work practices required by paragraph (f)(4) of this section.

(v) The contract employer shall advise the employer of any unique hazards presented by the contract employer's work, or of any hazards found during the contract employer's work."

Paragraphs (h)(3) (i) and (ii) require the communication of basic process hazard and emergency information by the contract employer to the contract employees. Paragraph (h)(3)(iii) requires the contract employer to document that each contract employee has received and understood the required training. Such a requirement is necessary to help ascertain that employees have been properly trained. Paragraph (h)(3)(iv) of the PSM Standard imposes a requirement pertaining to safe work practices contained in paragraph (f)(4). It is vitally important that contract employers assure that their employees follow the rules of the facility. Paragraph (h)(3)(v) requires that contract employers inform the plant employer of the hazards presented by the contractor's work, as well as any hazards found during the contractor's work.

5.5.1 Contractors: Compliance Guidelines

Employers who use contractors to perform work in and around processes that involve highly hazardous chemicals, will need to establish a screening process so that they hire and use contractors who accomplish the desired job tasks without compromising the safety and health of employees at a facility. For contractors, whose safety performance on the job is not known to the hiring employer, the employer will need to obtain information on injury and illness rates and experience and should obtain contractor references. Additionally, the employer must assure that the contractor has the appropriate job skills, knowledge and certifications (such as for pressure vessel welders). Contractor work methods and experiences should be evaluated. For example, does the contractor conducting demolition work swing loads over operating processes or does the contractor avoid such hazards?

Maintaining a site injury and illness log for contractors is another method employers must use to track and maintain current knowledge of work activities involving contract employees working on or adjacent to covered processes. Injury and illness logs of both the employer's employees and contract employees allow an employer to have full knowledge of process injury and illness experience. This log will also contain information which will be of use to those auditing process safety management compliance and those involved in incident investigations.

Contract employees must perform their work safely. Considering that contractors often perform very specialized and potentially hazardous tasks such as confined space entry activities and nonroutine repair activities it is quite important that their activities be controlled while they are working on or near a covered process. A permit system or work authorization system for these activities would also be helpful to all affected employers. The use of a work authorization system keeps an employer informed of contract employee activities, and as a benefit the employer will have better coordination and more management control over the work being performed in the process area. A well run and well maintained process where employee safety is fully recognized will benefit all of those who work in the facility whether they be contract employees or employees of the owner.

Chapter 5 - Training / Contractors 123

5.5.2 Contractors Compliance Checklist

1910.119 (h) CONTRACTORS

I. PROGRAM SUMMARY

The intent of this paragraph is to require employers who use contractors to perform work in and around processes that involve highly hazardous chemicals to establish a screening process so that they hire and use contractors who accomplish the desired job tasks without compromising the safety and health of employees at a facility. The contractor must assure that contract employees are trained on performing the job safely, of the hazards related to the job, and applicable provisions of the emergency action plan.

II. QUALITY CRITERIA REFERENCES

    A.   1910.119(h)
    B.   1910.119(f)(4)
    C.   1910.119(n)

III. VERIFICATION OF PROGRAM ELEMENTS	Criteria Reference	Met Y/N
A. Records Review - Employer's Program 1. Does the program include all contractors performing maintenance or repair, turnaround, major renovation or specialty work on or adjacent to covered processes? (Contractors performing incidental services which do not influence process safety such as janitorial work, food and drink services, laundry, delivery, and other supply services need not be included.) FIELD NOTE REFERENCE(S):	.119(h)(1)	
2. Is the information regarding the contractor's safety performance and programs obtained and evaluated for selection of contractors? FIELD NOTE REFERENCE(S):	.119(h)(2)(i)	

124 OSHA and EPA PSM Requirements

3. Are the contract employers informed of the known potential fire, explosion, or toxic release hazards related to the contractor's work and the processes? FIELD NOTE REFERENCE(S):	.119(h)(2)(ii)
4. Are contract employers informed of the applicable provisions of the emergency action plan required by .119(n)? FIELD NOTE REFERENCE(S):	.119(h)(2)(iii)
5. Have safe work practices to control the entrance, presence and exit of contract employers and contract employees in covered process areas been developed and implemented? FIELD NOTE REFERENCE(S):	.119(h)(2)(iv) .119(f)(4)
6. Are contract employers evaluated for their performance in fulfilling their obligations to: • Assure their employees are trained in safe work practices needed to perform the job? • Assure their employees are instructed in the known potential fire, explosion, or toxic release hazards related to the job and the applicable provisions of the emergency action plan? • Document the required training and the means to verify their employees have understood the training? • Assure their employees follow the facility safety rules and work practices? • Advise the employer of unique hazards presented by the contractor's work? FIELD NOTE REFERENCE(S):	.119(h)(2)(v)
7. Does the employer maintain a contract employee injury and illness log related to the contractor's work in process areas? FIELD NOTE REFERENCE(S):	.119(h)(2)(vi)

Records Review - Contractor's Programs

8. Are all contractor employees trained in the work practices necessary to perform their jobs safely? FIELD NOTE REFERENCE(S):	.119(h)(3)(i)

9. Is each contract employee instructed in the known potential fire, explosion, or toxic release hazards related to his/her job and the processes and applicable provisions of the emergency action plan? FIELD NOTE REFERENCE(S):	.119(h) (3)(ii)	
10. Is there documentation that each contract employee has received and understands the required training? FIELD NOTE REFERENCE(S):	.119(h) (3) (iii)	
11. Do the contract employee training records contain the following: • The identity of the employee? • The date of the training? • The means used to verify that the training was understood? FIELD NOTE REFERENCE(S):	.119(h) (3) (iii)	
12. Are there means to assure that contract employee follow the safety rules of the facility, including safe work practices required in .119(f)(4)? (Review evidence of enforcement by the contractor.) FIELD NOTE REFERENCE(S):	.119(h) (3)(iv) .119(f) (4)	
13. Is the employer advised of any unique hazards presented by the contract employer's work or any hazards found by the contract employer's work? FIELD NOTE REFERENCE(S):	.119(h) (3)(v)	
B. On-site Conditions 1. Based on a representative sample of observations of contractor employees, has the employer's program to control their entrance, presence, and exit been implemented? FIELD NOTE REFERENCE(S):	.119(h) (2)(iv)	

2. Based on a representative sample of observations of contractor employees, do they follow the safety rules of the facility? (These rules include the employer's safe work practices such as lockout/tagout, confined space entry, and opening process equipment or piping; they may also include other rules such as excavation procedures or use of PPE.) FIELD NOTE REFERENCE(S):	.119(h) (3)(iv)
C. Interviews 1. Based on interviews with contractor employers, have they been informed of the known fire, explosion, or toxic release hazards related to their work and the processes in which they are involved? FIELD NOTE REFERENCE(S):	.119(h) (2)(ii)
2. Based on interviews with contractor employers, have they been informed of the applicable provisions of the employer's emergency action plan? FIELD NOTE REFERENCE(S):	.119(h) (2) (iii)
3. Based on interviews with contractor employers and employees, have work practices to control their entrance, presence, and exit of covered process areas been implemented? FIELD NOTE REFERENCE(S):	.119(h) (2)(iv)
4. Based on interviews with the contractor employer, has the employer evaluated the contractor's performance in fulfilling the obligations required in .119(h)(3)? FIELD NOTE REFERENCE(S):	.119(h) (2)(v)
5. Based on interviews with a representative number of contractor employees, are they being instructed in the known potential fire, explosion, or toxic release hazards related to their work and the processes in which they are involved? FIELD NOTE REFERENCE(S):	.119(h) (3)(ii)

6. Based on interviews with a representative number of contractor employees, have they been instructed in the applicable provisions of the emergency action plan? (Ask them to explain the plan and evacuation procedures.) FIELD NOTE REFERENCE(S):	.119(h) (3)(ii)
7. Based on interviews with a representative number of contractor employees, has the contract employer assured that they follow the safety rules of the facility? (Ask how safe work practices, entry restrictions for the facility, and use of required PPE are enforced.) FIELD NOTE REFERENCE(S):	.119(h) (3)(iv)

5.5.3 Nonroutine Work Authorizations

Nonroutine work which is conducted in process areas needs to be controlled by the employer in a consistent manner. The hazards identified involving the work that is to be accomplished must be communicated to those doing the work, but also to those operating personnel whose work could affect the safety of the process. A work authorization notice or permit must have a procedure that describes the steps the maintenance supervisor, contractor representative or other person needs to follow to obtain the necessary clearance to get the job started. The work authorization procedures need to reference and coordinate, as applicable, lockout/tagout procedures, line breaking procedures, confined space entry procedures and hot work authorizations. This procedure also needs to provide clear steps to follow once the job is completed in order to provide closure for those that need to know the job is now completed and equipment can be returned to normal.

5.5.4 Contractor Training Settlement

Employers and the contractors they hire would both be responsible for ensuring that subcontractors are aware of and comply with the federal process safety standard for the chemical and petrochemical industries, according to terms of a settlement between the Occupational Safety and Health Administration and unions. (See "Contractor Training, Other Responsibilities Detailed in Settlement of Union Court Challenge," Occupational Safety & Health Daily (BNA) (April 16, 1993).)

The settlement clarifies the final agency standard following a court challenge from the United Steelworkers, the Oil, Chemical and Atomic Workers, and the Building Construction Trades Department of the AFL-CIO. The unions challenged the standard's effectiveness in protecting worker safety and health, particularly in provisions addressing safety training, employee involvement, and mandatory safety measures needed to prevent explosions and other hazards (*United Steelworkers v. OSHA*, 3rd Cir., No. 92-3106, petition filed 3/2/92).

Under the settlement, the process safety standard will be applied to all contractor activities involving process safety. Companies that hire contractors also must include in their periodic evaluations of contractor performance an assessment of training

Chapter 5 - Training / Contractors 129

provided to contractor employees. Both the primary, or direct-hire, employer and general contractors are responsible for informing subcontractors of potential hazards before work begins and ensuring compliance with the process safety standard, according to the agency. The primary employer also must consult with both its own employees and contract employees when developing the process safety hazard analyses required by the standard. Other settlement terms include requirements addressing maintenance worker training and the implementation of recommendations made by both process hazard analysis teams and incident investigation teams.

The clarifications will be included in the next update of OSHA Instruction CPL 2-2.45A, the compliance directive detailing enforcement interpretations of the process safety standard. (See **Appendix 9.**)

Chapter 6

Maintaining Mechanical Integrity and Managing Process Changes

6.1 MECHANICAL INTEGRITY OF EQUIPMENT

An active program of prevention and predictive maintenance is vital to maintain facilities with as-built integrity. Like incident investigations, assurance of mechanical integrity provides the opportunity to improve the reliability of emergency control and process equipment. Quality assurance addresses construction and maintenance of process safety equipment. It ensures that process equipment is fabricated in accordance with design specifications, assembled and installed properly, and maintained at its original level of integrity.

It is important to maintain the mechanical integrity of critical process equipment to ensure it is designed and installed correctly and operates properly. PSM mechanical integrity requirements apply to the following equipment:

*Pressure vessels and storage tanks;
*Piping systems (including piping components such as valves);
*Relief and vent systems and devices;
*Emergency shutdown systems;
*Controls (including monitoring devices and sensors, alarms, and interlocks); and
*Pumps.

The employer must establsh and implement written procedures to maintain the ongoing integrity of process equipment. Employees involved in maintaining the ongoing integrity of process equipment must be trained in an overview of that process and its hazards and trained in the procedures applicable to employee job tasks.

Inspection and testing must be performed on process equipment, using procedures that follow recognized and generally accepted good engineering practices. The frequency of inspections and tests of process equipment must conform with manufacturers' recommendations and good engineering practices, or more frequently if determined to be necessary by prior operating experience. Each inspection and test on process equipment must be documented, identifying the date of the inspection or test, the name of the person who performed the inspection or test, the serial number or other identifier of the equipment on which the inspection or test was performed, a description of the inspection or test performed, and the results of the inspection or test.

Equipment deficiencies outside the acceptable limits defined by the process safety information must be corrected before further use. In some cases, it may not be necessary that deficiencies be corrected before further use, as long as deficiencies are corrected in a safe and timely manner, when other necessary steps are taken to ensure safe operation.

In constructing new plants and equipment, the employer must ensure that equipment as it is fabricated is suitable for the process application for which it will be used. Appropriate checks and inspections must be performed to ensure that equipment is installed properly and is consistent with design specifications and the manufacturer's instructions. The employer also must ensure that maintenance materials, spare parts, and equipment are suitable for the process application for which they will be used.

6.2 MECHANICAL INTEGRITY: SPECIFIC REQUIREMENTS

Paragraph (j) of the PSM Standard contains the requirements for maintaining the mechanical integrity of process equipment in order to assure that such equipment is designed, installed, and operates properly. Paragraph (j)(1) specifies the process equipment to which the requirements of this paragraph apply. This equipment included pressure vessels and storage tanks; piping systems (including piping components such as valves); relief and vent systems and devices; emergency shutdown systems; and controls, alarms, and interlocks. OSHA felt that any of this equipment could have a significant impact on the safety of a process that is covered by this standard if the equipment was improperly designed or installed or, if such equipment did not function as intended.

It is OSHA's position that at least the equipment specified in paragraph (j)(1) must be subject to the requirements contained in paragraph (j). However, if an employer deems additional equipment to be critical to a particular process, that employer should consider that equipment to be covered by this paragraph and treat it accordingly.

Paragraph (j)(2) of the rule pertains to written procedures with respect to mechanical integrity. Paragraph (j)(2) requires the employer to establish and implement written procedures to maintain the on-going integrity of listed process equipment. The purpose of this provision is to require a written program that would ensure that process equipment receives careful, appropriate, regularly scheduled maintenance to ensure its continued safe operation.

Paragraph (j)(3) of the PSM Standard requires the employer to ensure that each employee involved in maintaining the on-going integrity of process equipment be trained in the procedures applicable to the employee's job tasks. OSHA was concerned that there might be some confusion between the training requirements in this mechanical integrity provision, and the training requirements contained in paragraph (g). It is the agency's position that maintenance employees need not be trained in process operating procedures to the same extent as those employees who are actually involved in operating the process. However, maintenance employees must receive on-going training in an overview of the process and its hazards and training in the procedures applicable to their job tasks to ensure that they can perform their tasks in a safe manner. Without continual attention to training needs due to process changes and other changes, little assurance will exist that maintenance employees will perform their tasks safely.

This provision has been assigned the title of "Training for process maintenance activities," and reads as follows:
"The employer shall train each employee involved in maintaining the on-going integrity of process equipment in an overview of that process and its hazards and in the procedures applicable to the employee's job tasks to assure that the employee can perform the job tasks in a safe manner."

Chapter 6 - Maintenance / Process Changes

Paragraph (j)(4) of the PSM Standard requires inspections and tests to be performed on specified process equipment because of the potential safety and health hazards that could result if the equipment malfunctioned. In an effort to assure that inspections and tests are performed properly, paragraph (j)(4)(ii) requires that inspection and test procedures follow generally accepted good engineering practices. OSHA's view is that "generally accepted good engineering practices" include appropriate internal standards of a facility, as well as codes and standards published by the American Society for Testing and Materials (ASTM), the American National Standards Institute (ANSI), the National Fire Protection Association (NFPA), and other organizations.

Paragraph (j)(4)(iii) of the PSM Standard requires the frequency of inspections and tests to be consistent with applicable manufacturers' recommendations and good engineering practices, and more frequently if determined necessary by prior operating experience. This provision is a performance-oriented requirement that provides flexibility for the employer to choose the frequency which will provide the best assurance of equipment integrity. OSHA's view is that the phrase "recognized and generally accepted good engineering practices" includes both appropriate internal standards and applicable codes and standards.

Paragraph (j)(4)(iv) requires the employer to document that each inspection and test has been performed in accordance with paragraph (j). It also requires that the documentation identify the date of the inspection; the name of the person who performed the inspection and test; and, the serial number or other identifier of the equipment.

Paragraph (j)(4)(iv) of the PSM Standard reads as follows:

"The employer shall document each inspection and test that has been performed on process equipment. The documentation shall identify the date of the inspection or test; the name of the person who performed the inspection or test; the serial number or other identifier of the equipment; the inspection or test that is performed; and, the results of the inspection or test."

Paragraph (j)(5) requires the employer to correct deficiencies in equipment which are outside acceptable limits before further use. The purpose of this provision is to require equipment deficiencies to be corrected promptly if the equipment is outside the acceptable limits specified in the process safety information. However, because there may be many situations where it may not be necessary that the deficiencies be corrected "before further use," OSHA also permits deficiencies to be corrected in a "safe and timely manner" as long as necessary means are taken to assure safe operation.

Consequently, paragraph (j)(5) of the rule reads as follows:

"The employer shall correct deficiencies in equipment that are outside the acceptable limits defined by the process safety information in paragraph (d) before further use, or in a safe and timely manner when necessary means are taken to assure safe operation."

Paragraph (j)(6) of the PSM Standard pertains to quality assurance of mechanical equipment. Paragraph (j)(6)(i) requires the employer to ensure that equipment as fabricated is suitable for the process application for which it will be used.

Paragraph (j)(6)(ii) requires appropriate checks and inspections to be performed as necessary to ensure that equipment is installed properly and consistent with design specifications and manufacturer's instructions. Paragraph (j)(6)(iii) requires the employer to ensure that maintenance materials, and spare parts and equipment are suitable for the process application for which they will be used.

6.2.1 Mechanical Integrity: Compliance Guidelines

Employers will need to review their maintenance programs and schedules to see if there are areas where "breakdown" maintenance is used rather than an on-going mechanical integrity program. Equipment used to process, store, or handle highly hazardous chemicals needs to be designed, constructed, installed and maintained to minimize the risk of releases of such chemicals. This requires that a mechanical integrity program be in place to assure the continued integrity of process equipment.

Elements of a mechanical integrity program include:

- [ ] Identification and categorization of equipment and instrumentation
- [ ] Inspections and tests
- [ ] Testing and inspection frequencies

Chapter 6 - Maintenance / Process Changes   135

[ ] Development of maintenance procedures
[ ] Training of maintenance personnel
[ ] Establishment of criteria for acceptable test results
[ ] Documentation of test and inspection results.
[ ] Documentation of manufacturer recommendations as to meantime to failure for equipment and instrumentation.

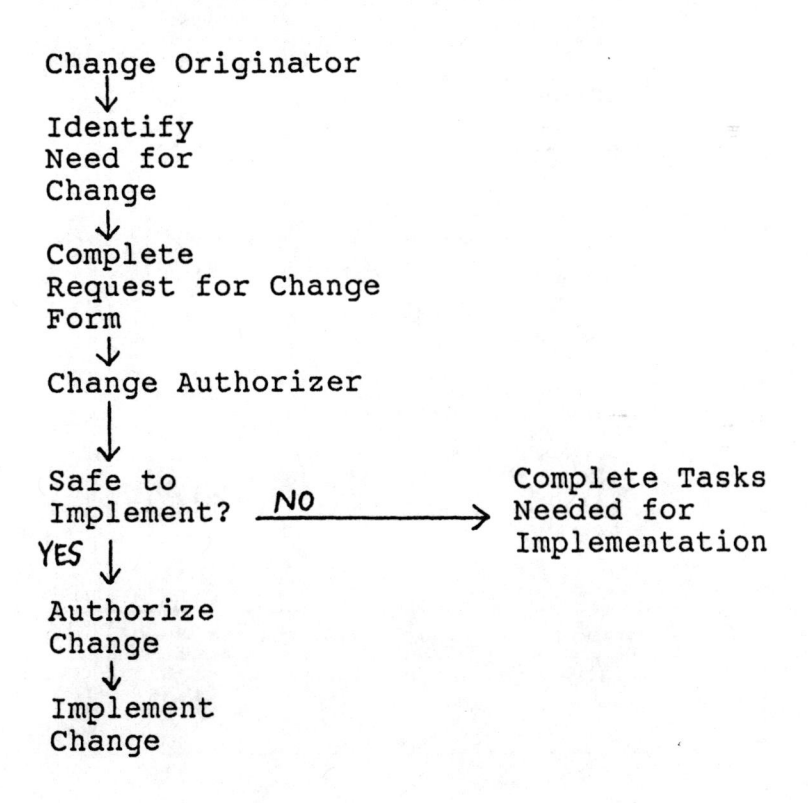

Figure 6A - Management of Change System

```
Change Originator
      ↓
Identify
Need for
Change
      ↓
Complete
Request for Change
Form
      ↓
Change Authorizer
      |
      ↓
Safe to                        Complete Tasks
Implement?   ──NO──→           Needed for
                               Implementation
YES ↓
Authorize
Change
      ↓
Implement
Change
```

6.2.2 Mechanical Integrity Compliance Checklist

1910.119 (j): MECHANICAL INTEGRITY

I. PROGRAM SUMMARY

The intent of this paragraph is to assure that equipment used to process, store, or handle highly hazardous chemical is designed, constructed, installed, and maintained to minimize the risk of releases of such chemicals. This requires that a mechanical integrity program be in place to assure the continued integrity of process equipment. The elements of a mechanical integrity program include the identification and categorization of equipment and instrumentation, development of written maintenance procedures, training for process maintenance activities, inspection and testing, correction of deficiencies in equipment that are outside acceptable limits defined by the process safety information, and development of a quality assurance program.

II. QUALITY CRITERIA REFERENCES

A. .119(j): Mechanical Integrity

III. VERIFICATION OF PROGRAM ELEMENTS	Criteria Reference	Met Y/N
A. Records Review 1. Does the written mechanical integrity program include? • Pressure vessels and storage tanks • Piping systems and components such as valves • Relief and vent systems and devices • Emergency shutdown systems • Controls (including monitoring devices and sensors, alarms and interlocks) • Pumps FIELD NOTE REFERENCE(S):	.119(j)(1)	
2. Are there written procedures to maintain the on-going integrity of process equipment? Does the documentation indicate the procedures have been implemented? FIELD NOTE REFERENCE(S):	.119(j)(2)	

Chapter 6 - Maintenance / Process Changes

3. Has training been provided to each employee involved in maintaining the on-going integrity of process equipment in the following: • An overview of the process and its hazards? • Procedures applicable to the employee's job tasks to assure that the employee can perform the job tasks in a safe manner? (Review certification documents for employees doing non-destructive tests, welding on pressure vessels, etc., where these certifications are required.) FIELD NOTE REFERENCE(S):	.119(j) (3)	
4. Are inspections and tests performed on each item of process equipment included in the program? FIELD NOTE REFERENCE(S):	.119(j) (4)(i)	
5. Do inspection and testing procedures follow good engineering practices? FIELD NOTE REFERENCE(S):	.119(j) (4)(ii)	
6. Are inspection and test frequencies consistent with the manufacturer's recommendation and good engineering practice? Are inspections and tests performed more frequently if determined necessary by operating experience? FIELD NOTE REFERENCE(S):	.119(j) (4) (iii)	
7. Is there documentation of each inspection and test that has been performed including all of the following: • Date of the inspection or test? • Name of person performing the procedure? • Serial number or other identifier of equipment on which procedure was performed? • Description of inspection or test performed? • Results of inspection or test? FIELD NOTE REFERENCE(S):	.119(j) (4)(iv)	

138 OSHA and EPA PSM Requirements

8. Are deficiencies in equipment that are outside limits (as defined in process safety information) corrected before further use or in a safe and timely manner when necessary means are taken to assure safe operation? FIELD NOTE REFERENCE(S):	.119(j)(5)	
9. In the construction of new plants and equipment, does the employer assure that equipment as it is fabricated is suitable for the process for which it will be used? FIELD NOTE REFERENCE(S):	.119(j)(6)(i)	
10. Have appropriate checks and inspections been made to assure equipment is installed properly and consistent with design specifications and manufacturer's instructions? (Include contractor supplied equipment.) FIELD NOTE REFERENCE(S):	.119(j)(6)(ii)	
11. Does the employer assure that maintenance materials, spare parts, and equipment are suitable for the process application for which they are used? (Include contractor supplied equipment.) FIELD NOTE REFERENCE(S):	.119(j)(6)(iii)	
B. On-site Conditions 1. Do observations of a representative sample of process equipment indicate deficiencies outside acceptable limits? (Compare process safety information criteria with the conditions of the equipment found in the process.) FIELD NOTE REFERENCE(S):	.119(j)(5)	
2. If new plants or equipment are being constructed, do observations indicate that the equipment as it is fabricated is suitable for the process application? FIELD NOTE REFERENCE(S):	.119(j)(6)(i)	

3. Do observations of a representative sample of maintenance materials, spare parts, and equipment indicate that they are suitable for the process application for which they will be used? FIELD NOTE REFERENCE(S):	.119(j) (6) (iii)	

C. Interviews

Engineers (if any; or other qualified persons capable of providing the information requested; see NOTE, p. A-2): 1. Based on interviews with a representative number of engineers, have procedures to maintain the on-going integrity of the process equipment been implemented for: • Pressure vessels and storage tanks? • Piping systems and components such as valves? • Relief and vent systems and devices? • Emergency shutdown systems? • Controls (including monitoring devices and sensors, alarms and interlocks)? • Pumps? (Ask about the possibility of safety critical equipment being inadvertently rendered inoperative. For example, a relief device might be isolated by closing an upstream valve.) FIELD NOTE REFERENCE(S):	.119(j) (2)	
2. Based on interviews with a representative number of engineers, do the inspection and testing procedures follow recognized and generally accepted good engineering practice? Has prior operating experience indicated a need for a more frequent test and inspection schedule than has been implemented? FIELD NOTE REFERENCE(S):	.119(j) (4)	
3. Based on interviews with a representative number of engineers, are equipment deficiencies corrected before use when they are outside the acceptable limits? If not, are the deficiencies corrected in a timely manner and are necessary means taken to assure safe operation? FIELD NOTE REFERENCE(S):	.119(j) (5)	

140 OSHA and EPA PSM Requirements

4. Based on interviews with a representative number of engineers, has the employer assured that, for new plants and equipment, the equipment as it is fabricated is suitable for the process application? Are appropriate checks and inspections made to assure equipment is installed properly and consistent with design specifications and manufacturer's instructions? Are maintenance materials, spare parts, and equipment suitable for the process application for which they will be used? (Ask about contractor supplied items.) FIELD NOTE REFERENCE(S):	.119(j)(6)	
Maintenance: 5. Based on interviews with a representative number of maintenance employees, have the written procedures for maintaining the on-going integrity of process equipment been implemented? FIELD NOTE REFERENCE(S):	.119(j)(2)	
6. Based on interviews with a representative number of employees involved in maintaining the on-going integrity of the process, have they been trained to assure they can perform their tasks in a safe manner? Did the training include an overview of the process, its hazards, and procedures applicable to the job? (Determine if certification, specialized training, or unique qualifications are required.) FIELD NOTE REFERENCE(S):	.119(j)(3)	
7. Based on interviews with a representative number of maintenance employees, do test and inspection procedures follow recognized and generally accepted good engineering practices? Is the frequency of inspections and tests consistent with applicable manufacturer's recommendations and good engineering practices? Are more frequent inspections and tests necessary due as indicated by prior operating experience? FIELD NOTE REFERENCE(S):	.119(j)(4)	

Chapter 6 - Maintenance / Process Changes 141

8. Based on interviews with a representative number of maintenance employees, are equipment deficiencies that are outside acceptable limits corrected before further use? If not, are corrections made in a timely manner and are necessary means taken to assure operation? FIELD NOTE REFERENCE(S):	.119(j) (5)	
9. Based on interviews with a representative number of maintenance employees, are maintenance materials, spare parts and equipment suitable for the process application for which they are intended? (Ask about availability and use of substitutes.) FIELD NOTE REFERENCE(S):	.119(j) (6)	

6.2.3 Lines of Defense

The first line of defense an employer has available is to operate and maintain the process as designed, and to keep the chemicals contained. This line of defense is backed up by the next line of defense which is the controlled release of chemicals through venting to scrubbers or flares, or to surge or overflow tanks which are designed to receive such chemicals, etc. These lines of defense are the primary lines of defense or means to prevent unwanted releases. The secondary lines of defense would include fixed fire protection systems like sprinklers, water spray, or deluge systems, monitor guns, etc., dikes, designed drainage systems, and other systems which would control or mitigate hazardous chemicals once an unwanted release occurs. These primary and secondary lines of defense are what the mechanical integrity program needs to protect and strengthen these primary and secondary lines of defense where appropriate.

The first step of an effective mechanical integrity program is to compile and categorize a list of process equipment and instrumentation for inclusion in the program. This list would include pressure vessels, storage tanks, process piping, relief and vent systems, fire protection system components, emergency shutdown systems and alarms and interlocks and pumps. For the categorization of instrumentation and the listed equipment the employer would prioritize which pieces of equipment require closer scrutiny than others. Meantime to failure of various instrumentation and equipment parts would be known from the manufacturers data or the employer's experience with the parts, which would then influence the inspection and testing frequency and associated procedures.

6.2.4 Inspection Methodologies

Applicable codes and standards such as the National Board Inspection Code, or those from the American Society for Testing and Material, American Petroleum Institute, National Fire Protection Association, American National Standards Institute, American Society of Mechanical Engineers, and other groups, provide information to help establish an effective testing and inspection frequency, as well as appropriate methodologies. The applicable codes and standards provide criteria for external inspections for such items as foundation and supports, anchor bolts, concrete or steel supports, guy wires, nozzles and sprinklers, pipe hangers, grounding connections, protective coatings and insulation, and external metal surfaces of piping and vessels, etc.

These codes and standards also provide information on methodologies for internal inspection, and a frequency formula based on the corrosion rate of the materials of construction. Also, both internal and external erosion needs to be considered along with corrosion effects for piping and valves. Where the corrosion rate is not known, a maximum inspection frequency is recommended, and methods of developing the corrosion rate are available in the codes. Internal inspections need to cover items such as vessel shell, bottom and head; metallic linings; nonmetallic linings; thickness measurements for vessels and piping; inspection for erosion, corrosion, cracking and bulges; internal equipment like trays, baffles, sensors and screens for erosion, corrosion or cracking and other deficiencies.

Some of these inspections may be performed by state or local government inspectors under state and local statutes. However, each employer needs to develop procedures to ensure that tests and inspections are conducted properly and that consistency is maintained even where different employees may be involved. Appropriate training is to be provided to maintenance personnel to ensure that they understand the preventive maintenance program procedures, safe practices, and the proper use and application of special equipment or unique tools that may be required. This training is part of the overall training program called for in the standard.

6.2.5 Quality Assurance System

A quality assurance system is needed to help ensure that the proper materials of construction are used, that fabrication and inspection procedures are proper, and that installation procedures recognize field installation concerns. The quality assurance program is an essential part of the mechanical integrity program and will help to maintain the primary and secondary lines of defense that have been designed into the process to prevent unwanted chemical releases or those which control or mitigate a release.

Too often commonplace and frequently inconspicuous hardware features of a plant design become the precursors of serious incidents. Inadequate attention is paid to seemingly insignificant plant features during design and construction phases, such as drains and vents, piping and equipment "pockets," small piping systems, and piping and equipment pressure rating specification breaks. These items must be closely scrutinized to ensure that they meet the necessary safety and operability requirements (Tomfohrde, 1985).

"As-built" drawings, together with certifications of coded vessels and other equipment, and materials of construction need to be verified and retained in the quality assurance documentation. Equipment installation jobs need to be properly inspected in the field for use of proper materials and procedures and to ensure that qualified craftsmen are used to do the job. The use of appropriate gaskets, packing, bolts, valves, lubricants and welding rods need to be verified in the field. Also procedures for installation of safety devices need to be verified, such as the torque on the bolts on ruptured disc installations, uniform torque on flange bolts, proper installation of pump seals, etc.

If the quality of parts is a problem, it may be appropriate to conduct audits of the equipment supplier's facilities to better assure proper purchases of required equipment which is suitable for its intended service. Any changes in equipment that may become necessary will need to go through the management of change procedures, discussed later in this chapter.

6.3 MANAGEMENT OF PROCESS CHANGES

Management of physical changes in the field ensures control over the everyday changes that occur in any process. These changes must be consistent with established process technology in order to avoid serious process incidents. Management must establish requirements for all field modifications prior to implementation of the modifications. Seemingly insignificant changes in the field can have far-reaching effects. For example, a worker repairs a piece of equipment and uses a different bolt. Later, different materials are used or process conditions changed to improve efficiency. Or perhaps the raw material supplier is replaced with another supplier. If there are no controls in place, changes can build upon changes until the equipment being operated is so far from the original design that the design review becomes obsolete.

Depending on the size, scope, and complexity of the plant processes, the plant management should develop objectives and procedures for management of change. Management of change should be keyed to local plant needs. Management of change objectives may be organized into a formal design specification for a management system. Such a design specification should communicate the concerns and expectations of plant management by outlining several key features, including: terminology, roles and responsibilities, the scope of the management of change system, interface with other plant practices and programs, guidelines for management of change issues, and requirements for review and authorization of changes.

6.3.1 Developing a Management of Change System

Once company management has established a design specification for a management of change system, an appropriate interdisciplinary team should be assembled to develop a written management of change program. The written management of change program will be used to educate and train plant personnel in specific management of change procedures. The following checklist outlines tasks that must be completed to develop the management of change system:

*Identify potential change situations

146 OSHA and EPA PSM Requirements

*Coordinate the management of change system with existing plant procedures
*Establish Request-For-Change review and approval procedures
*Develop guidelines for key management of changes issues
*Design management of change system documentation
*Define employee training requirements
*Establish procedures for modifiying the management of change system
*Determine whether the management of change system meets the design specification

Figure 6A is a flowchart for a basic management of change system.

6.4 MANAGEMENT OF CHANGE: PSM STANDARD REQUIREMENTS

Contemplated changes to a process must be thoroughly evaluated to fully assess their impact on employee safety and health and to determine needed changes to operating procedures. To this end, the PSM Standard contains a section on procedures for managing changes to processes. Written procedures to manage changes (except for "replacements in kind") to process chemicals, technology, equipment, and procedures, and change to facilities that affect a covered process, must be established and implemented. These written procedures must ensure that the following considerations are addressed prior to any change:

*Technical basis for the proposed change;
*Impact of the change on employee safety and health;
*Modifications to operating procedures;
*Necessary time period for the change; and
*Authorization requirements for the proposed change.

Employees who operate a process and maintenance and contract employees whose job tasks will be affected by a change in the process must be informed of, and trained in, the change prior to startup of the affected part of the process. If a change covered by these results in a change in the required process safety information, such information also must be updated accordingly. If a change covered by these procedures changes the required operating procedures or practices, they also must be updated.

One of the most important and necessary aspects of a process safety management program is appropriately managing changes to the process. Many of the incidents investigated by OSHA resulted from some type of change to the process (See examples discussed in Chapters 1 and 7).

Paragraph (l) of the PSM Standard addresses management of change. Paragraph (l)(1) requires the employer to establish and implement written procedures to manage changes (except for "replacement in kind") to process chemicals, technology, equipment and procedures; and changes to facilities that affect a covered process.

Paragraph (l)(2) requires that the following considerations be addressed prior to any change:

148 OSHA and EPA PSM Requirements

(i) The technical basis for the proposed change;
(ii) Impact of change on safety and health;
(iii) Modifications to operating procedures;
(iv) Necessary time period for the change; and
(v) Authorization requirements for the proposed change.

Paragraph (1)(3) requires that employees involved in the process be informed of and trained in the change in the process as early as practicable prior to its startup. This information and training provision applies to operating employees as well as to maintenance and contract employees whose job tasks will be affected by the change. Otherwise, contract employees or maintenance employees who are unaware of the change, may unwittingly cause an incident by doing their job tasks as they have in the past.

Paragraph (1)(3) of the rule reads as follows:

"Employees involved in operating a process and maintenance and contract employees whose job tasks will be affected by a change in the process shall be informed of, and trained in, the change prior to start-up of the process or the affected part of the process."

Paragraph (1)(4) of the rule requires that if a change covered by this provision results in a change to the process safety information, that such information be appended and/or updated in accordance with paragraph (d) of the PSM rule. Finally, paragraph (1)(5) requires that if a change covered by this provision results in a change to the operating procedures, such procedures must be appended and/or updated in accordance with paragraph (f) of the PSM rule.

6.4.1 Managing Change: Compliance Guidelines

To properly manage changes to process chemicals, technology, equipment and facilities, one must define what is meant by change. In the process safety management standard, change includes all modifications to equipment, procedures, raw materials and processing conditions other than "replacement in kind". These changes need to be properly managed by identifying and reviewing them prior to implementation of the change. For example, the operating procedures contain the operating parameters (pressure limits, temperature ranges, flow rates, etc.) and the importance of operating within these limits. While the operator must have the flexibility to maintain safe operation within the established parameters, any operation outside of these parameters requires review and approval by a written management of change procedure.

Management of change covers changes in process technology and changes to equipment and instrumentation. Changes in process technology can result from changes in production rates, raw materials, experimentation, equipment unavailability, new equipment, new product development, change in catalyst and changes in operating conditions to improve yield or quality. Equipment changes include, among others, change in materials of construction, equipment specifications, piping pre-arrangements, experimental equipment, computer program revisions and changes in alarms and interlocks. Employers need to establish means and methods to detect both technical changes and mechanical changes.

Temporary changes have caused a number of catastrophes over the years, and employers need to establish ways to detect temporary changes as well as those that are permanent. It is important that a time limit for temporary changes be established and monitored since, without control, these changes may tend to become permanent. Temporary changes are subject to the management of change provisions. In addition, the management of change procedures are used to ensure that the equipment and procedures are returned to their original or designed conditions at the end of the temporary change. Proper documentation and review of these changes are invaluable in assuring that the safety and health considerations are being incorporated into the operating procedures and the process.

150 OSHA and EPA PSM Requirements

6.4.2 Management of Change Compliance Checklist

1910.119 (l): MANAGEMENT OF CHANGE

I. PROGRAM SUMMARY

The intent of this paragraph is to require management of all modifications to equipment, procedures, raw materials and processing conditions other than "replacement in kind" by identifying and reviewing them prior to implementation of the change. Minimum requirements for management of change include: establishing written procedures to manage change; addressing the technical basis, impact on safety and health, modification to operating procedures, necessary time period, and authorizations required; informing and training employees affected; and updating process safety information and operating procedures or practices.

II. QUALITY CRITERIA REFERENCES

A. 1910.119 (l): Management of Change

III. VERIFICATION OF PROGRAM ELEMENTS	Criteria Reference	Met Y/N
A. Records Review 1. Are there written procedures for managing changes (except for "replacements in kind") to process chemicals, technology, equipment, and procedures and changes to facilities that affect a covered process? (Review procedures that address responsibilities, steps for assessing risks and approving changes, requirements for reviewing designs for temporary and permanent changes, steps needed to verify that modifications have been made as designed, variance procedures, time limit authorizations for temporary changes, and steps required to return the process to status quo after temporary changes.) FIELD NOTE REFERENCE(S):	.119(l)(1)	
2. Do the procedures assure that the technical basis for the proposed change is addressed prior to any change? FIELD NOTE REFERENCE(S):	.119(l)(2)(i)	

Chapter 6 - Maintenance / Process Changes

3. Do the procedures assure that the impact of the change on safety and health is addressed prior to any change? FIELD NOTE REFERENCE(S):	.119(1) (2)(ii)
4. Do the procedures assure that modifications to operating procedures is addressed prior to any change? FIELD NOTE REFERENCE(S):	.119(1) (2) (iii)
5. Do the procedures assure that the necessary time period for the change is addressed prior to any change? FIELD NOTE REFERENCE(S):	.119(1) (2)(iv)
6. Do the procedures assure that the authorization requirements for the proposed change are addressed prior to any change? FIELD NOTE REFERENCE(S):	.119(1) (2)(v)
7. Are employees involved in operating a process, and maintenance and contract employees whose job tasks will be affected by change informed of, and trained in, the change prior to start-up of process or affected part of process? FIELD NOTE REFERENCE(S):	.119(1) (3)
8. Is the process safety information required by paragraph (d) updated if changed? FIELD NOTE REFERENCE(S):	.119(1) (4)
9. Are the operating procedures or practices required by paragraph (f) updated if changed? FIELD NOTE REFERENCE(S):	.119(1) (5)

B. On-site Conditions

1. Do observations of new or recently modified process chemicals, technology, equipment, or procedures (except "replacement in kind") indicate that the Management of Change procedures have been implemented?

(Determine if records are available to support the procedures for new or revised processes found in the facility.)

FIELD NOTE REFERENCE(S):

.119(l)(1)

C. Interviews

Operators, Maintenance, and Contractor Employees:

1. Based on interviews with operators, maintenance employees and contractor employees, are procedures implemented to manage changes to existing process chemicals, technology, equipment, facilities, and procedures?

FIELD NOTE REFERENCE(S):

.119(l)(1)

2. Based on interviews with operators, maintenance employees and contractor employees, is training in process changes provided to employees whose job tasks will be affected by the changes prior to start-up?

FIELD NOTE REFERENCE(S):

.119(l)(3)

6.4.3 Request for Change Form

Employers may wish to develop a form or clearance sheet to facilitate the processing of changes through the management of change procedures. A typical change form may include a description and the purpose of the change, the technical basis for the change, safety and health considerations, documentation of changes for the operating procedures, maintenance procedures, inspection and testing, P&IDs, electrical classification, training and communications, pre-startup inspection, duration if a temporary change, approvals and authorization.

Where the impact of the change is minor and well understood, a checklist reviewed by an authorized person with proper communication to others who are affected may be sufficient. However, for a more complex or significant design change, a hazard evaluation procedure with approvals by operations, maintenance, and safety departments may be appropriate. **Figure 6B** provides an example of a general request for change form.

Changes in documents such as P&IDs, raw materials, operating procedures, mechanical integrity programs, electrical classifications, etc., need to be noted so that these revisions can be made permanent when the drawings and procedure manuals are updated. Copies of process changes need to be kept in an accessible location to ensure that design changes are available to operating personnel as well as to PHA team members when a PHA is being done or one is being updated.

154 OSHA and EPA PSM Requirements

6.4.4 Sample Request for Change Form

```
Unit or Area _____    RFC No. _____
                                     Date    _____
DESCRIPTION AND REASON FOR CHANGE:




_____
Originator
================================================================
[ ]  Temporary Change (If change is temporary, list
     pertinent dates)  Dates Valid: _____
================================================================
Environmental, health, and safety reviews are complete
and all concerns have been addressed.



_____
EH&S Review Team Leader
================================================================
Operating, maintenance, and emergency procedures have
been reviewed.



_____
Area Procedures Coordinator
================================================================
All affected personnel have been informed of the change.
The appropriate training has been given.



_____
Area Training Coordinator
================================================================
All affected process safety information is scheduled for
revision.



_____
Unit Engineer
================================================================
[ ] PSSR Required          PSSR No. _____
================================================================
This change has met the appropriate review requirements
and has been approved.



_____
Area Manager
```

Chapter 6 - Maintenance / Process Changes   155

6.5 PRE-STARTUP SAFETY REVIEWS

Major changes in a process often require that Pre-Startup Safety Reviews (PSSRs) be conducted before the process returns to operation. The PSSR confirms that the equipment is in accordance with design specifications, appropriate procedures are in place, and necessary training is given. Traditionally, management of change procedures addressed changes in, or minor additions to, existing facilities. PSSRs were conducted for new facilities, restart of suspended operations, and extensive modifications that required shutdown of existing facilities, modification, and restart. This approach treated management of change and PSSR programs separately. However, in complying with the OSHA PSM Standard, a facility must have closely related programs for management of change and PSSR. Therefore, to coordinate management of change and PSSR procedures, a company may wish to develop a single procedure that describes both of them. Alternatively, if separate procedures are maintained for each, a company should ensure that the separate management of change and PSSR procedures cross-reference to one another, explaining how they work together to review process changes and maintain regulatory compliance.

6.5.1 Pre-startup Safety Review: PSM Standard Requirements

It is important that a safety review takes place before any highly hazardous chemical is introduced into a process. The PSM Standard, therefore, requires the employer to perform a pre-startup review for new facilities and for modified facilities when the modification is significant enough to require a change in the process safety information. Prior to the introduction of a highly hazardous chemical to a process, the pre-startup safety review must confirm the following:

*Construction and equipment are in accordance with design specifications;
*Safety, operating, maintenance, and emergency procedures are in place and are adequate;
*A process hazard analysis has been performed for new facilities and recommendations have been resolved or implemented before startup, and modified facilities meet the management of change requirements; and

*Training of each employee in operating a process has been completed.

Paragraph (i)(1) requires the employee to perform a pre-startup safety review for new facilities and for modified facilities when the modification necessitates a change to the process safety information. The purpose of this requirement is to make sure that certain important considerations have been addressed before any highly hazardous chemical is introduced into a process.

It was not the intent of OSHA to require a pre-startup safety review for each facility that may be modified slightly. OSHA's position is that a pre-startup safety review is necessary for modified facilities only when the modification is significant enough to require a change in the process safety information. Proposed paragraph (i)(1) has been revised to read as follows:

> "The employer shall perform a pre-startup safety review for new facilities and for modified facilities when the modification is significant enough to require a change in the process safety information."

Paragraph (i)(2) of the PSM Standard requires that the pre-startup safety review confirm that construction is in accordance with design specifications ((i)(2)(i)); safety, operating, maintenance, and emergency procedures are in place and were adequate ((i)(2)(ii)); process hazard analysis recommendations have been addressed and actions necessary for startup have been completed ((i)(2)(iii)); and, operating procedures are in place and training of each operating employee has been completed ((i)(2)(iv)). Paragraph (i)(2) of the PSM Standard reads as follows:

> "(2) The pre-startup safety review shall confirm that prior to the introduction of highly hazardous chemicals to a process:
> (i) Construction and equipment is in accordance with design specifications;
> (ii) Safety, operating, maintenance, and emergency procedures are in place and are adequate;
> (iii) For new facilities, a process hazard analysis has been performed and recommendations have been resolved or implemented before startup; and modified facilities

meet the requirements contained in management of change, paragraph (l).
(iv) Training of each employee involved in operating a process has been completed."

6.5.2 Pre-Startup Safety: Compliance Guidelines

For new processes, the employer will find a PHA helpful in improving the design and construction of the process from a reliability and quality point of view. The safe operation of the new process will be enhanced by making use of the PHA recommendations before final installations are completed. P&IDs are to be completed along with having the operating procedures in place and the operating staff trained to run the process before startup. The initial startup procedures and normal operating procedures need to be fully evaluated as part of the pre-startup review to assure a safe transfer into the normal operating mode for meeting the process parameters.

For existing processes that have been shutdown for turnaround, or modification, etc., the employer must assure that any changes other than "replacement in kind" made to the process during shutdown go through the management of change procedures. P&IDs will need to be updated as necessary, as well as operating procedures and instructions. If the changes made to the process during shutdown are significant and impact the training program, then operating personnel as well as employees engaged in routine and nonroutine work in the process area may need some refresher or additional training in light of the changes. Any incident investigation recommendations, compliance audits or PHA recommendations need to be reviewed as well to see what impacts they may have on the process before beginning the startup.

6.5.3 Pre-Startup Safety Review Compliance Checklist

1910.119 (i): PRE-STARTUP SAFETY REVIEW

I. PROGRAM SUMMARY

The intent of this paragraph is to make sure that, for new facilities and for modified facilities when the modification necessitates a change to process safety information, certain important considerations are addressed before any highly hazardous chemicals are introduced into the process. Minimum requirements include that the pre-startup safety review confirm the following: construction and equipment is in accordance with design specifications; safety, operating, maintenance, and emergency procedures are in place and adequate; for new facilities, a PHA has been performed and recommendations resolved or implemented; modified facilities meet the requirements of paragraph (l), management of change; and training of each employee involved in the process has been completed.

II. QUALITY CRITERIA REFERENCES

    A.   1910.119(i): Pre-startup Safety Review
    B.   1910.119(l): Management of Change

III. VERIFICATION OF PROGRAM ELEMENTS	Criteria Reference	Met Y/N
A. Records Review 1. Has a pre-startup safety review been performed for all new facilities and for modified facilities when the modification is significant enough to require a change in process safety information? FIELD NOTE REFERENCE(S):	.119(i) (1)	
2. Do pre-startup safety reviews confirm that prior to the introduction of highly hazardous chemicals to a process: • Construction and equipment is in accordance with design specifications? FIELD NOTE REFERENCE(S): • Safety, operating, maintenance, and emergency procedures are in place and adequate? FIELD NOTE REFERENCE(S):	.119(i) (2)	

Chapter 6 - Maintenance / Process Changes

2. (Continued) Do pre-startup safety reviews confirm that prior to the introduction of highly hazardous chemicals to a process: • For new facilities, a PHA has been performed and recommendations resolved or implemented before startup? FIELD NOTE REFERENCE(S):	.119(i)(2)
• Modified facilities meet requirements of paragraph (l)? FIELD NOTE REFERENCE(S):	
• Training of each employee involved in operating the process has been completed? FIELD NOTE REFERENCE(S):	
On-site Conditions 1. Do observations of new or modified facilities indicate that prior to the introduction of highly hazardous chemicals: • Construction and equipment is in accordance with design specifications? • Safety, operating, maintenance, and emergency procedures are in place and adequate? FIELD NOTE REFERENCE(S):	.119(i)(2)
C. Interviews [See NOTE, p. A-2.] 1. Based on interviews with a representative sample of operators, maintenance employees, and engineers, can it be confirmed that the construction and equipment are in accordance with design specifications prior to introducing highly hazardous chemicals to a process? FIELD NOTE REFERENCE(S):	.119(i)2(i)
2. Based on interviews with a representative sample of operators, maintenance employees, and engineers, are safety, operating, maintenance, and emergency procedures in place prior to introduction of highly hazardous chemicals into a process? Are these procedures adequate? FIELD NOTE REFERENCE(S):	.119(i)2(ii)

160 OSHA and EPA PSM Requirements

3. Based on interviews with a representative sample of operators, maintenance employees, and engineers, is a PHA performed and are recommendations resolved prior to a startup that introduces highly hazardous chemicals into a new process? FIELD NOTE REFERENCE(S):	.119(i) 2(iii)	
4. Based on interviews with a representative sample of operators, maintenance employees, and engineers, do modified facilities meet requirements of paragraph (1), Management of Change prior to introducing a highly hazardous chemical. FIELD NOTE REFERENCE(S):	.119(i) 2(iii)	
5. Based on interviews with a representative sample of operators, is training completed for each employee involved in operating the process prior to the introduction of a highly hazardous chemical? FIELD NOTE REFERENCE(S):	.119(i) 2(iv)	

Chapter 7

Incident Investigation, Emergency Planning, and Compliance Audits

7.1 INVESTIGATION AND RESPONSE TO PROCESS INCIDENTS

Management must determine the causes of all serious and potentially serious incidents and take steps to prevent recurrence. Management must plan ahead to deal with incidents effectively. Even the best-run sites sometimes experience incidents, but with an investigation procedure in place, management will be prepared to handle them and learn from them. (See Section 7.1.1.)

Management must develop emergency planning and response procedures to minimize the impact of any process incidents that occur and to bring emergency situations under prompt control. As a result of community right-to-know laws, and as a good business practice generally, industry's focus must extend beyond the property line and into surrounding communities. Industry needs emergency response plans and procedures to help protect the public and to demonstrate to the public that it effectively controls its processes.

7.1.1 Case Histories of Process Incidents

Flixborough, England, 28 fatalities (1974): Fire & Explosion
This accident occurred because a bypass assembly, a piece of 20" pipe, was not properly analyzed for stress nor properly supported. When the plant bypassed a disabled reactor, the bypass assembly failed and released a large vapor cloud of cyclohexane which subsequently ignited.

Seveso, Italy widespread contamination (1976): Dioxin release
This accident occurred when an exothermic decomposition reaction caused the rupture disc on the reactor to fail, venting dioxin into the atmosphere and contaminating the surrounding countryside. In this case, the production run in question ended at 6 a.m. Saturday morning, a time which coincided with the closing of the plant for the weekend. The release from the exothermic reaction took place some 6 1/2 hours later. A number of procedural violations occurred in order to terminate the production run by 6 a.m. Saturday, including:

>*Only 15 percent of the solvent was distilled off at the end of the batch versus 50 percent as specified by operating procedures.
>*No water was added to cool the reactor to 50-60 degrees centigrade versus the 3,000 liters specified.
>*Stirring stopped after 15 minutes instead of stirring until fully cooled as specified by the operating procedure.
>*Operators left at 6 a.m. instead of remaining with the unit until cooling to 50-60 degrees centigrade as specified by the operating procedure.

Bhopal, India, 2,000+ fatalities (1984): Methyl isocyanate release
This accident is believed to have occurred when water entered a storage tank containing methyl isocyanate. At the time of the incident, the following emergency control equipment was either not in commission or not in full working order:

>*Vent gas scrubber designed to scrub vapor discharged from the Methyl isocyanate storage tank relief valves.
>*Flare system designed to burn vapor which gets through the vent gas scrubber.
>*Cooling system for the storage tank.

Flame arrester incident (Howard, 1988)
A major chemical company had a large partial oxidation unit. The desired product was recovered in an absorber to protect the environment, the off-gas containing traces of organics was then sent to an incinerator. If a certain kind of realistically possible upset occurred, a flammable mixture of hydrocarbon gas plus air could go through the process equipment to the incinerator. To prevent flashback of the flame, from the incinerator into the

Chapter 7 - Incidents / Emergency Planning / Audits   163

upstream equipment, the engineering contractor installed what is called a flame arrester.

For the 4 ft.diameter pipeline, the flame arrester was 12 ft. in diameter, and the arrester element consisted of a 3 ft. thick, 12 ft. diameter section of 2 in. Pall rings. The total included angle of inlet and outlet cones was about 150 degrees.

Two inch Pall rings will not stop the combustion of hydrocarbon-air mixtures that are reasonably within the flammable range. The technology of dependably functional flame arresters is controlled by the fundamental characteristics of flames and by fundamentals of how a flame arrester stops a flame, which is not merely by heat transfer. Thus, although the large arrester looked impressive in its elevated position in the equipment support structure, it was nonfunctional. The inevitable happened. A process upset caused flashback from the incinerator, and the flame arrester failed to stop the flame. The engineering design of the flame arrester unit was obviously defective.

An actual failure during an incident is not needed to uncover such a faulty design. The chemical company could identify the problem during process hazard reviews, and then seek help to correct the design error.

Need for hard-wired backup of critical safety instrumentation (Howard, 1988)

A small but technically-oriented chemical company wanted to install a new process that involved many significant hazards because of the possibility of exothermic runaway reactions and the severe toxicity of materials in the process. These hazards, in turn, pointed to the need for the extensive use of process safety interlocks and alarms. The chemical company hired a major engineering firm to design and construct the new process unit, and the contractor presented a computerized system that completely controlled the operation and included all the necessary safety interlocks and alarms. The chemical company was impressed with the total computer control system, which would be the first for the company.

The contractor did not present certain critical process safety concepts to the chemical company, concepts that the chemical company did not self-generate internally. Critical safety interlocks and alarms must have a backup. For example, controlling the level of critical liquid may require a "high-level" interlock and alarm plus a "high-high-level" interlock and alarm.

In general, the two critical sensing devices should not be of the same type. What is needed is not mere redundancy but diversity. Instead of running both sets of safety devices through a computer, the critical backup devices should be hard-wired. When the chemical company learned about these concepts during a hazard review, it was able to readily incorporate them into the project.

Inadequate venting of exothermic runaway reactions (Howard, 1988)

Numerous major incidents - many involving explosions, and some with loss of life - have occurred because venting of exothermic runaway reactions was inadequate. One company conducted numerous batch organic chemical reactions in its plant. The reactors used ranged in size from approximately 25 to 1,000 gallons. Through the years the plant followed the custom of installing a 2 inch relief vent line on each reactor. Each relief line had a rupture disc that was usually set to burst near the pressure rating of each vessel, which normally was about 100 p.s.i.g. A number of organic chemical reactions were either highly exothermic themselves or could undergo a highly exothermic runaway reaction at some stage.

Even though a 2 in. vent line could be grossly inadequate for venting reaction runaways in vessels above a given size, the engineers were so accustomed to the vent sizing practice that they never questioned the adequacy of the vent sizes or the relief pressure of the rupture discs. In recent years, the bases for vent sizing have improved significantly over previously used methods. Companies must develop an understanding of this technology and implement it to properly safeguard against improper venting of exothermic reactions.

7.2 INCIDENT INVESTIGATION

A crucial component of every process safety management program is thorough investigation of incidents to identify the chain of events and causes so that corrective measures can be developed and implemented. Accordingly, the PSM Standard requires the investigation of each incident that resulted in, or could reasonably have resulted in, a catastrophic release of a highly hazardous chemical in the workplace.

Such an investigation must be initiated as promptly as possible, but not later than 48 hours following the incident. The

Chapter 7 - Incidents / Emergency Planning / Audits 165

investigation must be completed by a team consisting of at least one person knowledgeable in the process involved, including a contract employee if the incident involved the work of a contractor, and other persons with appropriate knowledge and experience to investigate and analyze the incident thoroughly.

An investigation report must be prepared with coverage of at least the following items:

*Date of the incident;
*Date the investigation began;
*Description of the incident;
*Factors that contributed to the incident; and
*Recommendations resulting from the investigation

A system must be established to promptly address and resolve the incident report findings and recommendations. Resolutions and corrective actions must be documented and the report reviewed by all affected personnel whose job tasks are relevant to the incident findings (including contract employees when applicable). The employer must retain these incident investigation reports for at least five years.

7.2.1 Incident Investigation: Specific Requirements

OSHA included requirements for incident investigation in the PSM Standard because a crucial part of any process safety management program is the thorough investigation of any incident that resulted in, or could reasonably have resulted in, a catastrophic release of a highly hazardous chemical in the workplace. Such investigations are extremely important for identifying the chain of events leading to the incident and for determining causal factors. Information resulting from the investigation will be invaluable to the development and implementation of corrective measures and for use in subsequent process hazard analyses.

Paragraph (m)(1) requires the employer to investigate every incident which results in, or could reasonably have resulted in (near miss), a catastrophic release in the workplace. Paragraph (m)(1) reads as follows:

"The employer shall investigate each incident which resulted in, or could reasonably have resulted in, a catastrophic release of a highly hazardous chemical in the workplace."

Paragraph (m)(2) requires incident investigations to be initiated as promptly as possible, but no later than 48 hours following the incident. It is important that an incident investigation be initiated promptly so that events can be recounted as clearly as possible; to preserve crucial evidence; and so that there is less likelihood that the scene will have been disturbed. Still, OSHA realized that circumstances may not facilitate an immediate investigation because of the potential emergency nature of some incidents. Thus, this provision allows enough flexibility to the employer by requiring an incident investigation be initiated as soon as possible but not later than 48 hours following the incident. OSHA felt that 48 hours was a reasonable timeframe within which to initiate an investigation. It should also be noted that the investigation need only be initiated within this timeframe, not completed, although it is contemplated that there will not be unnecessary delay between initiation and completion of the incident investigation.

Paragraph (m)(3) of the PSM Standard requires that an incident investigation team be established and consist of persons knowledgeable in the process involved and other appropriate specialties, as necessary. OSHA does not require an employee representative to be on the process hazard analysis team or on the incident investigation team.

The intent of OSHA is to ensure that team members have the ability to properly perform the investigation promptly and that the employer have the flexibility to select team members (in consultation with employees and their representatives as described in paragraph (c) of the rule) that possess this ability. Additionally, the agency felt that in cases where an incident involves a contract employer's work, then a contract employee should be involved in the investigation. Therefore, paragraph (m)(3) reads as follows:

> "An incident investigation team shall be established and consist of at least one person knowledgeable in the process involved, including a contract employee if the incident involved work of the contractor, and other persons with appropriate knowledge and experience to thoroughly investigate and analyze the incident."

Paragraph (m)(4) requires a report to be prepared at the conclusion of the investigation which includes, at a minimum, the date of the incident; date that the investigation began; a description of the incident; the factors that contributed to the incident; and, any recommendations resulting from the investigation. OSHA wanted to make sure that the investigation is initiated promptly. Consequently, it is important that the date of the incident, as well as the date that the investigation was initiated, are both specified.

Paragraph (m)(6) requires the employer to establish a system to promptly address the report findings and recommendations and to implement the report recommendations in a timely manner. This provision also requires that resolution and corrective action be documented.

Paragraph (m)(6) requires that the report be reviewed with all affected personnel whose job tasks are relevant to the incident findings, including contract employees when applicable. The purpose of this provision is to ensure that the report findings are disseminated to appropriate personnel, because the information contained in the report might be important in preventing similar incidents.

Paragraph (m)(7) of the PSM Standard requires incident investigation reports to be retained for five years in order to determine if an incident pattern develops or exists. OSHA considered a three-year retention period. However, the agency felt that it would be extremely useful if the report findings and recommendations were reviewed during the subsequent update or revalidation of the process hazard analysis. Consequently, the agency decided it was more appropriate to specify a five-year retention period to be consistent with paragraph (e) of the PSM rule, which requires the process hazard analysis to be updated or revalidated every five years.

7.2.2 Investigation of Incidents: Compliance Guidelines

Incident investigation is the process of identifying the underlying causes of incidents and implementing steps to prevent similar events from occurring. The intent of an incident investigation is for employers to learn from past experiences and thus avoid repeating past mistakes. The incidents for which OSHA expects employers to become aware of and to investigate are the types of events which result in, or could reasonably have resulted in, a catastrophic release. Some of the events are sometimes referred to as "near misses," meaning that a serious consequence did not occur, but could have.

Employers need to develop in-house capability to investigate incidents that occur in their facilities. A team needs to be assembled by the employer and trained in the techniques of investigation including how to conduct interviews of witnesses, needed documentation and report writing. A multi-disciplinary team is better able to gather the facts of the event and to analyze them and develop plausible scenarios as to what happened, and why. Team members should be selected on the basis of their training, knowledge and ability to contribute to a team effort to fully investigate the incident. Employees in the process area where the incident occurred should be consulted, interviewed or made members of the team. Their knowledge of the events form a significant set of facts about the incident which occurred. The report, its findings and recommendations are to be shared with those who can benefit from the information. The cooperation of employees is essential to an effective incident investigation. The focus of the investigation should be to obtain facts, and not to place blame. The team and the investigation process should clearly deal with all involved individuals in a fair, open and consistent manner. **Figure 7A** is a Sample Investigation Report.

Chapter 7 - Incidents / Emergency Planning / Audits 169

Figure 7A - Sample Investigation Report

National Safety Council

ACCIDENT INVESTIGATION REPORT — CASE NUMBER

COMPANY _____ ADDRESS _____

DEPARTMENT _____ LOCATION (if different from mailing address) _____

1 NAME of INJURED	2 SOCIAL SECURITY NUMBER	3 SEX ☐ M ☐ F	4 AGE	5 DATE of ACCIDENT
6 HOME ADDRESS	7 EMPLOYEE'S USUAL OCCUPATION		8 OCCUPATION at TIME of ACCIDENT	

	9 LENGTH of EMPLOYMENT	10 TIME in OCCUP at TIME of ACCIDENT
11 EMPLOYMENT CATEGORY ☐ Regular, full-time ☐ Temporary ☐ Nonemployee ☐ Regular, part-time ☐ Seasonal	☐ Less than 1 mo ☐ 6 mos to 5 yrs ☐ 1-5 mos ☐ More than 5 yrs	☐ Less than 1 mo ☐ 6 mos to 5 yrs ☐ 1-5 mos ☐ More than 5 yrs
13 NATURE of INJURY and PART of BODY	12 CASE NUMBERS and NAMES of OTHERS INJURED in SAME ACCIDENT	

14 NAME and ADDRESS of PHYSICIAN	16 TIME of INJURY A _____ AM / PM B Time within shift C Type of shift	17 SEVERITY of INJURY ☐ Fatality ☐ Lost workdays—days away from work ☐ Lost workdays—days of restricted activity ☐ Medical treatment ☐ First aid ☐ Other, specify _____
15 NAME and ADDRESS of HOSPITAL		

18 SPECIFIC LOCATION of ACCIDENT ON EMPLOYER'S PREMISES? ☐ Yes ☐ No	19 PHASE OF EMPLOYEE's WORKDAY at TIME of INJURY ☐ During rest period ☐ Entering or leaving plant ☐ During meal period ☐ Performing work duties ☐ Working overtime ☐ Other _____

20 DESCRIBE HOW the ACCIDENT OCCURRED

_____
_____
_____
_____

21. ACCIDENT SEQUENCE Describe in reverse order of occurrence events preceding the injury and accident Starting with the injury and moving backward in time, reconstruct the sequence of events that led to the injury

A Injury Event _____
B Accident Event _____
C Preceding Event #1 _____
D Preceding Event #2, #3, etc _____

170 OSHA and EPA PSM Requirements

22 TASK and ACTIVITY at TIME of ACCIDENT	23 POSTURE of EMPLOYEE
A General type of task _____	_____
B Specific activity _____	24 SUPERVISION at TIME of ACCIDENT
C Employee was working	☐ Directly supervised ☐ Not supervised
☐ Alone ☐ With crew or fellow worker ☐ Other, specify _____	☐ Indirectly supervised ☐ Supervision not feasible

25 CAUSAL FACTORS Events and conditions that contributed to the accident. Include those identified by use of the Guide for Identifying Causal Factors and Corrective Actions

_____
_____
_____
_____
_____
_____
_____
_____
_____
_____
_____
_____

26 CORRECTIVE ACTIONS Those that have been, or will be, taken to prevent recurrence. Include those identified by use of the Guide for Identifying Causal Factors and Corrective Actions

_____
_____
_____
_____
_____
_____
_____
_____
_____
_____
_____
_____
_____

PREPARED BY _____	APPROVED _____
TITLE _____	TITLE _____ DATE _____
DEPARTMENT _____ DATE _____	APPROVED _____
Reprinted with permission from the National Safety Council, 1121 Spring Lake Drive, Itasca, IL 60143-3201	TITLE _____ DATE _____

Chapter 7 - Incidents / Emergency Planning / Audits 171

7.2.3 Incident Investigations Compliance Checklist

1910.119 (m): INCIDENT INVESTIGATIONS

I. PROGRAM SUMMARY

The employer is required to investigate each incident which resulted in, or could reasonably have resulted in a catastrophic release of highly hazardous chemical in the workplace. An investigation shall be initiated no later than 48 hours following the incident. An investigation team shall be established and a report prepared which includes: 1) Date of incident 2) Date investigation began 3) Description of incident 4) Factors that contributed to the incident 5) Recommendations from the investigation. The employer is required to establish a system to promptly address the incident report findings and recommendations, documenting all resolutions and corrective actions. Incident reports shall be reviewed with all affected personnel whose job tasks are relevant to the investigation and retained for five years.

II. QUALITY CRITERIA REFERENCES

A. 1910.119(m): Incident Investigations

III. VERIFICATION OF PROGRAM ELEMENTS	Criteria Reference	Met Y/N
A. Records Review		
1. Has each incident been investigated which resulted in, or could reasonably have resulted in a catastrophic release of highly hazardous chemicals in the workplace? FIELD NOTE REFERENCE(S):	.119(m) (1)	
2. Have incident investigations been initiated as promptly as possible, but not later than 48 hours following the incident? FIELD NOTE REFERENCE(S):	.119(m) (2)	

3. Have incident investigation teams been established? Do the teams contain at least one person knowledgeable in the process involved in the incident, and other members with appropriate knowledge and experience to thoroughly investigate and analyze the incident? Has a contractor employee been included in the team if the incident involved work of the contractor? FIELD NOTE REFERENCE(S):	.119(m) (3)	
4. Have incident investigation reports been prepared at the conclusion of the investigation which include at minimum: • Date of the incident? • Date the inspection began? • A description of the incident? • The factors that contributed to the incident? • Any recommendations resulting from the investigation? FIELD NOTE REFERENCE(S):	.119(m) (4)	
5. Has a system been established to promptly address and resolve the incident investigation report findings and recommendations? FIELD NOTE REFERENCE(S):	.119(m) (5)	
6. Have resolutions and corrective actions from the incident investigation reports been documented? FIELD NOTE REFERENCE(S):	.119(m) (5)	
7. Have incident investigation reports been reviewed with all affected personnel whose job tasks are relevant to the incident findings including contract employees, where applicable? FIELD NOTE REFERENCE(S):	.119)m) (6)	
8. Are incident investigation reports retained for five years? FIELD NOTE REFERENCE(S):	.119(m) (7)	

B. On-site Conditions

1. Do observations of a representative sample of process components involved in incident investigations indicate that recommendations have been resolved?

(Compare the corrective actions outlined in the investigation documentation with the actual equipment, procedures, material use, etc.)

FIELD NOTE REFERENCE(S):

.119(m)(5)

C. Interviews

1. Based on interviews with a representative number of operators, maintenance employees and contractor employees, have all incidents that resulted in or could reasonably have resulted in a catastrophic release of highly hazardous chemicals in the workplace, been investigated?

FIELD NOTE REFERENCE(S):

.119(m)(1)

2. Based on interviews with a representative number of the members of past investigation teams, do the teams contain at least one person knowledgeable in the process involved in the incident, and other persons with appropriate knowledge and experience to thoroughly investigate and analyze the incident? Was a contractor employee included in the team if the incident involved work of the contractor?

FIELD NOTE REFERENCE(S):

.119(m)(3)

3. Based on interviews with a representative number of employees whose job tasks are relevant to the past incident investigation findings, have the investigation reports been reviewed with the affected personnel?

FIELD NOTE REFERENCE(S):

.119(m)(6)

7.3 EMERGENCY PLANNING AND RESPONSE: SPECIFIC REQUIREMENTS

If, despite the best planning, an incident occurs, it is essential that emergency pre-planning and training make employees aware of, and able to execute, proper actions. For this reason, an emergency action plan for the entire plant must be developed and implemented in accordance with the provisions of other OSHA rules (e.g., 29 CFR 1910.38(a)). In addition, the emergency action plan must include procedures for handling small releases of hazardous chemicals. Employers covered under the PSM Standard may be subject to the OSHA hazardous waste and emergency response regulations (29 CFR 1910.120(a), (p), and (q)).

Paragraph (n) of the PSM rule requires the employer to establish and implement an emergency action plan in accordance with the provisions contained in 29 CFR 1910.38(a). For information purposes the Agency also added a note that 29 CFR 1910.120 (a), (p) and (q) may also be applicable. Implementation of an emergency action plan is extremely important for plant sites which have processes covered by this standard because of the potential hazards posed by highly hazardous chemicals and the elements of the emergency action plan which must be implemented to pre-plan for emergencies involving these substances (including training) so that employees will be aware of, and execute, appropriate actions.

The emergency action plan requires, at a minimum, the implementation of, and training employees in, the following procedures:

[ ] Emergency escape procedures and emergency escape route assignments.
[ ] Procedures to be followed by employees who remain to operate critical plant operations before they evacuate.
[ ] Procedures to account for all employees after emergency evacuation has been completed.
[ ] Rescue and medical duties for those employees who are to perform them.
[ ] Preferred means of reporting fires and other emergencies.

Chapter 7 - Incidents / Emergency Planning / Audits

[ ] Names or regular job titles of persons who can be contacted for further information or explanation of duties under the plan.

The emergency action plan also requires the establishment of a system to alert employees of an emergency. If the alarm system is to be used for alerting fire brigade members, or for some other purpose, a distinctive signal must be used for each purpose. With respect to training, employers must review the emergency action plan with each employee initially when the plan is developed, whenever the employee's responsibilities or designated actions under the emergency action plan changes, and whenever the emergency action plan itself is changed.

The pre-planning and training required by the emergency action plan will ensure the readiness of employees to respond appropriately and safely to emergencies involving highly hazardous chemicals. Additionally, as a part of emergency planning, employers must develop procedures to address small releases and spills since it is not always obvious when such an event is, or is not, an emergency situation. Such an event may also warrant initiating an incident investigation.

OSHA concluded that drills or simulated exercises are also recommended but should not be mandatory. Thus, no specific provision was included in the PSM rule to require drills. OSHA felt that the employer is in the best position to assess the readiness of employees to respond correctly, to establish procedures for emergency action, including conducting drills or exercises when necessary. Additionally, the subject of drills is adequately addressed by the elements contained in the emergency action plan and applicable provisions of 29 CFR 1910.120.

7.3.1 Emergency Preparedness: Compliance Guidelines

Each employer must address what actions employees are to take when there is an unwanted release of highly hazardous chemicals. Emergency preparedness or the employer's tertiary (third) lines of defense are those that will be relied on along with the secondary lines of defense when the primary lines of defense which are used to prevent an unwanted release fail to stop the release. Employers will need to decide whether: (1) they want employees to handle and stop small or minor incidental releases; or (2) they wish to mobilize the available resources at the plant and have them brought to bear on a more significant release; or (3) they want employees to evacuate the danger area and promptly escape to a preplanned safe zone area, and allow the local community emergency response organizations to handle the release; or (4) they want to use some combination of these actions. Employers will need to select how many different emergency preparedness or tertiary lines of defense they plan to have, develop the necessary plans and procedures, appropriately train employees in their emergency duties and responsibilities, and then implement these lines of defense.

Employers, at a minimum, must have an emergency action plan which will facilitate the prompt evacuation of employees due to an unwanted release of a highly hazardous chemical. This means that the employer will have a plan that will be activated by an alarm system to alert employees when to evacuate and, that employees who are physically impaired, will have the necessary support and assistance to get them to the safe zone as well. The intent of these requirements is to alert and move employees to a safe zone quickly. Delaying alarms or confusing alarms are to be avoided. The use of process control centers or similar process buildings in the process area as safe areas is discouraged. Recent catastrophes have shown that a large loss of life has occurred in these structures because of where they have been sited and because they are not necessarily designed to withstand overpressures from shockwaves resulting from explosions in the process area.

Unwanted incidental releases of highly hazardous chemicals in the process area must be addressed by the employer so that employees know what actions to take. If the employer wants employees to evacuate the area, then the emergency action plan will be activated. For outdoor processes where wind direction is important for selecting the safe route to a refuge area, the employer should place a wind direction indicator such as a wind sock or pennant at the highest point that can be seen throughout the process area. Employees can move in the direction of crosswind to upwind to gain safe access to the refuge area by knowing the wind direction.

If the employer wants specific employees in the release area to control or stop the minor emergency or incidental release, these actions must be planned for in advance and procedures developed and implemented. Pre-planning for handling incidental releases for minor emergencies in the process area needs to be

Chapter 7 - Incidents / Emergency Planning / Audits

done, appropriate equipment for the hazards must be provided, and training needs to be conducted for those employees who will perform the emergency work before they respond to handle an actual release. The employer's training program, including the Hazard Communication Standard training is to address the training needs for employees who are expected to handle incidental or minor releases.

7.3.2 Emergency Planning and Response Compliance Checklist

1910.119 (n): EMERGENCY PLANNING AND RESPONSE

I. PROGRAM SUMMARY

The intent of this paragraph is to require the employer to address what actions employees are to take when there is an unwanted release of highly hazardous chemicals. The employer must establish and implement an emergency action plan in accordance with the provisions of 29 CFR 1910.38(a) and include procedures for handling small releases. Certain provisions of the hazardous waste and emergency response standard, 29 CFR 1910.120(a), (p), and (q), may also apply.

[NOTE: 1910.120(a) addresses scope, application, and definitions for the entire standard. 1910.120(p) addresses treatment, storage, and disposal (TSD) facilities under the Resource Conservation and Recovery Act (RCRA). 1910.120(q) addresses requirements for facilities that are not RCRA TSD's, where there is the potential for an emergency incident involving hazardous substances. Cleanup operations--including corrective actions and post-emergency response cleanup--are covered by 1910.120(b) through (o). For further guidance, refer to the forthcoming directive on 29 CFR 1910.120.]

II. QUALITY CRITERIA REFERENCES

A.	1910.119(n)	D.	1910.165
B.	1910.38(a)	E.	1910.1200
C.	1910.120(a),(p),(q)	F.	1910.36(b)

III. VERIFICATION OF PROGRAM ELEMENTS

	Criteria Reference	Met Y/N

A. Records Review

1. Has an emergency action plan been established and implemented for the entire plant in accordance with the requirements of 1910.38? Does the plan address the following:
• Escape procedures and routes?
• Procedures for post-evacuation employee accounting?
• Preferred means to report emergencies?
• Duties and procedures of employees who:
  • Remain to operate critical equipment?
  • Perform rescue and medical duties?
• The names for persons or locations to contact for more action plan information?
• Employee alarm systems?
 	
FIELD NOTE REFERENCE(S):
	.119(n) or .38(a)(2)	

2. Is the plan written if the facility has more than ten employees? FIELD NOTE REFERENCE(S):	.38(a)(1)	
3. Is there sufficient number of persons designated and trained to assist in the safe and orderly emergency evacuation of employees? FIELD NOTE REFERENCE(S):	.38(a)(5)(i)	
4. Is the plan reviewed with each employee covered by the plan: initially when the plan is developed; and whenever the employees' responsibilities or designated action under the plan change; and whenever the plan is changed? FIELD NOTE REFERENCE(S):	.38(a)(5)(ii)	
5. Does the emergency action plan cover procedures for handling small releases? FIELD NOTE REFERENCE(S):	.119(n)	

180 OSHA and EPA PSM Requirements

6. Is an alarm system established and implemented which complies with 1910.165? Are the alarms:
 	
• Distinctive for each purpose of the alarm?
 	
• Capable of being perceived above ambient noise and light levels by all employees in the affected portions of the workplace?
 	
• Distinctive and recognizable as a signal to evacuate the work area or perform actions designated under the plan?
 	
• Maintained in operating condition?
 	
• Tested appropriately and restored to normal operating condition as soon as possible after test?
   • Non-supervised systems tested not less than every two months?
   • Supervised systems tested at least annually?
 	
• Serviced, maintained, and tested by appropriately trained persons?
 	
• Unobstructed, conspicuous and readily accessible, if they are manual alarm systems?
 	
FIELD NOTE REFERENCE(S):
	.165 (b-e)	

7. Does the written emergency response plan meet the requirements of 1910.120 (a), (p), and (q), if appropriate? (See the NOTE at I., Program Summary. Clean-up operations required by a governmental body are addressed in .120(a); treatment, storage, and disposal (TSD) facilities under the Resource Conservation and Recovery Act are addressed in .120(p); and .120(q) addresses requirements for emergency response no matter where they occur, except that it does not cover employees engaged in operations at TSD facilities or hazardous waste sites.) FIELD NOTE REFERENCE(S):	.120(a) .120(p) .120(q)	

8. If employees are engaged in emergency response (except clean-up operations), does the plan address the following: • Coordination with outside parties? • Personnel roles, lines of authority, training, and communication? • Emergency recognition and prevention? • Safe distances and places of refuge? • Site security and control? • Evacuation routes and procedures? • Decontamination? • Emergency medical treatment and first aid? • Emergency alerting and response procedures? • Critique of response and followup? • PPE and emergency equipment? FIELD NOTE REFERENCE(S):	.120(q)	

B. On-site Conditions

1. Do observations of a representative sample of alarm systems indicate that they comply with the requirements in .165(b-e)? Are the alarms: • Distinctive for each purpose of the alarm? • Capable of being perceived above ambient noise and -light levels by all employees in the affected portions of the workplace? • Distinctive and recognizable as a signal to evacuate the work area or perform actions designated under the plan? • Maintained in operating condition? • Tested appropriately and restored to normal operating condition as soon as possible after test? (Be present for an alarm test if possible or review video if available.) • Tested no greater than every two months? • Serviced, maintained, and tested by appropriately trained persons? • Unobstructed, conspicuous and readily accessible, if they are manual alarm systems? FIELD NOTE REFERENCE(S):	.165 (b-e)	
2. Do observations of the evacuation routes indicate that they are not blocked, locked, or barricaded? FIELD NOTE REFERENCE(S):	.36(b) (4)	

3. Do observations of the evacuation routes indicate that there are readily visible signs for evacuation routes leading to safe locations? FIELD NOTE REFERENCE(S):	.36(b)(5)	
4. Do observations of a representative sample of the evacuation route signs during dark conditions indicate that they are adequately illuminated? FIELD NOTE REFERENCE(S):	.36(b)(6)	
## C. Interviews 1. Based on interviews with employees who have been identified as likely to discover releases or assigned other emergency response duties, are they provided training? Is the training based on the duties they are expected to perform? FIELD NOTE REFERENCE(S):	.120(q)(6)	
2. Based on interviews with employees who are likely to discover hazardous substance releases, can they demonstrate competency in the provisions listed in the first responder awareness level: • Understanding what hazardous substances are, and the risks associated with them in an incident? • Understanding potential outcomes associated with an emergency when hazardous substances are present? • Ability to recognize the presence of hazardous substances in an emergency? • Ability to identify the hazardous substances, if possible? • Understanding the role of the first responder awareness individual in the employer's emergency response plan, including site security and control and the U.S. Dept. of Transportation's Emergency Response Guidebook? • Ability to realize the need for additional resources, and make appropriate notifications to the communication center? FIELD NOTE REFERENCE(S):	.120(q)(6)(i)	

3. Based on interviews with employees who will take defensive action in containing and controlling a release as part of the response, can they demonstrate the competencies for a first responder operations level: • Knowledge of the basic hazard and risk assessment techniques? • Knowledge of how to select and use proper PPE provided to them? • Understanding of basic hazardous materials terms? • Knowledge of how to perform basic containment, confinement, and control operations within the capability of their unit? • Knowledge of how to implement basic decontamination procedures? • Knowledge of relevant standard operating procedures and termination procedures for a response? FIELD NOTE REFERENCE(S):	.120(q) (6)(ii)	
4. Based on interviews with employees who will take offensive action in containing and controlling a release as part of the response, can they demonstrate the competencies for a hazardous materials (HAZMAT) technician: • Knowledge of how to implement the employer's emergency response plan? • Knowledge of the classification, identification, and verification of known and unknown materials using field survey instruments and equipment? • Ability to function within an assigned role in the Incident Command System? • Knowledge of how to select and use proper specialized chemical PPE provided to them? • Understanding of hazard and risk assessment techniques? • Ability to perform advanced control, containment, and/or confinement operations within the capability of their unit? • Understanding of how to implement decontamination procedures? • Understanding of termination procedures? • Understanding of basic chemical and toxicological terminology and behavior? FIELD NOTE REFERENCE(S):	.120(q) (6) (iii)	
5. Based on interviews with a representative number of operator and maintenance employees, do they know the emergency action plan to protect themselves in an emergency? FIELD NOTE REFERENCE(S):	.38(a)	

184 OSHA and EPA PSM Requirements

7.3.3 Pre-Planning for Releases

Pre-planning for releases that are more serious than incidental releases is another important line of defense to be used by the employer. When a serious release of a highly hazardous chemical occurs, the employer will have determined in advance, through pre-planning, what actions employees are to take. The evacuation of the immediate release area and other areas as necessary would be accomplished under the emergency action plan. If the employer wishes to use plant personnel such as a fire brigade, spill control team, a hazardous materials team, or use employees to render aid to those in the immediate release area and control or mitigate the incident, these actions are covered by 29 CFR 1910.120, the Hazardous Waste Operations and Emergency Response (HAZWOPER) Standard. If outside assistance is necessary, such as through mutual aid agreements between employers or local government emergency response organizations, these emergency responders are also covered by HAZWOPER. The safety and health protections required for emergency responders are the responsibility of their employers and of the on-scene incident commander.

7.3.4 Plant and Local Community Coordination of Emergency Response

Responders may be working under very hazardous conditions and therefore the objective is to have them competently led by an on-scene incident commander and the commander's staff, properly equipped to do their assigned work safely, and fully trained to carry out their duties safely before they respond to an emergency. Drills, training exercises, or simulations with the local community emergency response planners and responder organizations is one means to obtain better preparedness. This close cooperation and coordination between plant and local community emergency preparedness managers will also aid the employer in complying with the Environmental Protection Agency's Risk Management Plan criteria discussed in **Chapters 8 through 10**.

One effective way for medium to large facilities to enhance coordination and communication during emergencies between plant operations and local community organizations is for employers to establish and equip an emergency control center. The

Chapter 7 - Incidents / Emergency Planning / Audits   185

emergency control center would be sited in a safe zone area so that it could be occupied throughout the duration of an emergency. The center would serve as the major communication link between the on-scene incident commander and plant or corporate management as well as with the local community officials. The communication equipment in the emergency control center should include a network to receive and transmit information by telephone, radio or other means. It is important to have a backup communication network in case of power failure or one communication means fails. The center should also be equipped with the plant layout and community maps, utility drawings including fire water, emergency lighting, appropriate reference materials such as a government agency notification list, company personnel phone list, SARA Title III reports and material safety data sheets, emergency plans and procedures manual, a listing with the location of emergency response equipment, mutual aid information, and access to meteorological or weather condition data and any dispersion modeling data.

7.4   PSM COMPLIANCE AUDITS

Auditing is a crucial part of every successful management effort. Observing personnel at work and comparing actual performance to established standards are the only ways management can know and understand what is really happening at the worksite. Proper auditing requires positive feedback on significant strengths, as well as corrective feedback on areas needing improvement. Management must establish a regular schedule for periodic auditing and must ensure that auditors report their results for analysis. It is not enough to establish and implement a process safety management program and assume it will function. Even the most comprehensive program will erode with time if it is not audited and revised.

7.4.1   Compliance Audits: Specific Requirements

Paragraph (o) of the PSM Standard contains provisions pertaining to an evaluation of an employer's process safety management system. An audit with respect to compliance with the provisions contained in this section is an extremely important function. This is because it serves as a self-evaluation for employers to measure the effectiveness of their process safety

186 OSHA and EPA PSM Requirements

management system. The audit can identify problem areas and assist employers in directing attention to process safety management weaknesses. Therefore, paragraph (o)(1) requires employers to certify that they have evaluated compliance with the provisions of this section, at least every three years.

The objective of paragraph (o) is to ensure that employers evaluate the effectiveness of their process safety management system as required by the PSM Standard. An effective means of achieving this objective is by employers ensuring that the provisions contained in the PSM Standard are being met and, in doing so, the employer will ascertain whether the procedures and practices required to be developed under the process safety management standard as adequate and being followed.

Paragraph (o)(2) requires that at least one person knowledgeable in the process conduct the compliance audit. Although it is important for the audit to be performed by at least one person knowledgeable in the process, OSHA decided it was not necessary that it be performed by a team. Paragraph (o)(3) requires a report of the findings of the audit to be developed.

Paragraph (o)(4) requires the employer to promptly determine and document an appropriate response to each of the findings of the compliance audit, and document that deficiencies have been corrected. The purpose of this proposed paragraph is to assure that employers determine an appropriate response to each of the report findings and if employers identify a deficiency that needs to be corrected, that they "document" the correction of the deficiency.

Paragraph (o)(5) requires employers to retain the two most recent compliance audit reports, as well as the documented actions described in paragraph (o)(4) of this provision. The purpose of this proposed provision is to focus on any continuing areas of concern that are identified through the compliance audits.

7.4.2 Compliance Audits: PSM Standard Compliance Guidelines

Employers need to select a trained individual or assemble a trained team of people to audit the process safety management system and program. A small process or plant may need only one knowledgeable person to conduct an audit. The audit is to include an evaluation of the design and effectiveness of the process safety management system and a field inspection of the safety and health conditions and practices to verify that the employer's systems are effectively implemented. The audit should be conducted or led by a person knowledgeable in audit techniques and who is impartial toward the facility or area being audited. The essential elements of an audit program include planning, staffing, conducting the audit, evaluation and corrective action, follow-up and documentation. Planning in advance is essential to the success of the auditing process. Each employer needs to establish the format, staffing, scheduling and verification methods prior to conducting the audit. The format should be designed to provide the lead auditor with a procedure or checklist which details the requirements of each section of the standard. The names of the audit team members should be listed as part of the format as well. The checklist, if properly designed, could serve as the verification sheet which provides the auditor with the necessary information to expedite the review and assure that no requirements of the standard are omitted. This verification sheet format could also identify those elements that will require evaluation or a response to correct deficiencies. This sheet could also be used for developing the follow-up and documentation requirements.

188 OSHA and EPA PSM Requirements

7.4.3 Compliance Audits: Compliance Checklist

1910.119 (o): COMPLIANCE AUDITS

I. PROGRAM SUMMARY

The intent of this paragraph is to require employers to self-evaluate the effectiveness of their PSM program by identifying deficiencies and assuring corrective actions. Minimum requirements include: audits at least every three years; maintenance of audit reports for at least the last two audits; audits conducted by at least one person knowledgeable in the process; documentation of an appropriate response to each finding; documentation that the deficiencies found have been corrected.

II. QUALITY CRITERIA REFERENCES

    A.   1910.119(o): Compliance Audits
    B.   1910.119(c): Employee Participation

III. VERIFICATION OF PROGRAM ELEMENTS

	Criteria Reference	Met Y/N
A. Records Review		
1. Has the employer certified in writing that there has been an audit of compliance with PSM at least every three years? FIELD NOTE REFERENCE(S):	.119(o)(1)	
2. Do the audit reports include an evaluation of all the required paragraphs of the PSM standard? FIELD NOTE REFERENCE(S):	.119(o)(1)	
3. Was the compliance audit conducted by at least one person who was knowledgeable in the process? FIELD NOTE REFERENCE(S):	.119(o)(2)	
4. Has a report of the findings been developed for each audit? FIELD NOTE REFERENCE(S):	.119(o)(3)	

Chapter 7 - Incidents / Emergency Planning / Audits

5. Has the employer promptly determined and documented an appropriate response to each of the findings? FIELD NOTE REFERENCE(S):	.119(o) (4)
6. Does the employer document that deficiencies have been corrected? FIELD NOTE REFERENCE(S):	.119(o) (4)
7. Has the employer retained the two most recent compliance audit reports? FIELD NOTE REFERENCE(S):	.119(o) (5)
B. On-site Conditions No observations are required; on-site conditions will be cited under other paragraphs.	.119(o) (4)
C. Interviews 1. Based on interviews with auditors, are they knowledgeable in processes? FIELD NOTE REFERENCE(S):	.119(o) (2)
2. Based on interviews with a representative number of employees and their designated representatives, do they have access to compliance audit information? FIELD NOTE REFERENCE(S):	.119(c) (3)

7.4.4 Audit Team

The selection of effective audit team members is critical to the success of the program. Team members should be chosen for their experience, knowledge, and training and should be familiar with the processes and with auditing techniques, practices and procedures. The size of the team will vary depending on the size and complexity of the process under consideration. For a large, complex, highly instrumented plant, it may be desirable to have team members with expertise in process engineering and design, process chemistry, instrumentation and computer controls, electrical hazards and classifications, safety and health disciplines, maintenance, emergency preparedness, warehousing or shipping, and process safety auditing. The team may use part-time members to provide for the depth of expertise required as well as for what is actually done or followed, compared to what is written.

7.4.5 Review of Relevant Documentation

An effective audit includes a review of the relevant documentation and process safety information, inspection of the physical facilities, and interviews with all levels of plant personnel. Utilizing the audit procedure and checklist developed in the pre-planning stage, the audit team can systematically analyze compliance with the provisions of the standard and any other corporate policies that are relevant. For example, the audit team will review all aspects of the training program as part of the overall audit. The team will review the written training program for adequacy of content, frequency of training, effectiveness of training in terms of its goals and objectives as well as to how it fits into meeting the standard's requirements, documentation, etc. Through interviews, the team can determine the employee's knowledge and awareness of the safety procedures, duties, rules, emergency response assignments, etc. During the inspection, the team can observe actual practices such as safety and health policies, procedures, and work authorization practices. This approach enables the team to identify deficiencies and determine where corrective actions or improvements are necessary.

An audit is a technique used to gather sufficient facts and information, including statistical information, to verify compliance with standards. Auditors should select as part of their pre-planning a sample size sufficient to give a degree of confidence

Chapter 7 - Incidents / Emergency Planning / Audits

that the audit reflects the level of compliance with the standard. The audit team, through this systematic analysis, should document areas which require corrective action as well as those areas where the process safety management system is effective and working in an effective manner. This provides a record of the audit procedures and findings, and serves as a baseline of operation data for future audits. It will assist future auditors in determining changes or trends from previous audits.

7.4.6 Corrective Action

Corrective action is one of the most important parts of the audit. It includes not only addressing the identified deficiencies, but also planning, follow-up, and documentation. The corrective action process normally begins with a management review of the audit findings. The purpose of this review is to determine what actions are appropriate, and to establish priorities, timetables, resource allocations and requirements and responsibilities. In some cases, corrective action may involve a simple change in procedure or minor maintenance effort to remedy the concern. Management of change procedures need to be used, as appropriate, even for what may seem to be a minor change. Many of the deficiencies can be acted on promptly, while some may require engineering studies or in-depth review of actual procedures and practices. There may be instances where no action is necessary and this is a valid response to an audit finding. All actions taken, including an explanation where no action is taken on a finding, needs to be documented as to what was done and why.

It is important to assure that each deficiency identified is addressed, the corrective action to be taken noted, and the audit person or team responsible be properly documented by the employer. To control the corrective action process, the employer should consider the use of a tracking system. This tracking system might include periodic status reports shared with affected levels of management, specific reports such as completion of an engineering study, and a final implementation report to provide closure for audit findings that have been through management of change, if appropriate, and then shared with affected employees and management. This type of tracking system provides the employer with the status of the corrective action. It also provides the documentation required to verify that appropriate corrective actions were taken on deficiencies identified in the audit.

7.4.7 Preparing for the Audit

At the company level, the audit is planned by the managers from operations, technical, engineering, and safety. A date compatible with plant personnel availability is selected for the audit. Depending on the size and scope of the plant operations, the audit takes from several days to several weeks. The audit team, including plant managers and appropriate staff members, meet initially to discuss the scope of the audit as well as likely exposure areas or other areas of concern. The plant manager suggests the audit role, which encompasses all plant processes, operations, and facilities. When the audit team reaches a specific area, either during the review and discussion of processes and materials or site visits, each team member takes notes of observed hazards. The unit or areas noted during the audit that pose a hazard or have had a significant loss or a near-miss, should be subjected to a full process safety review (Krivan, 1986).

7.4.8 Checklist of Items to Examine During the Audit

[ ] Written instructions for the operation in place
[ ] Written instructions current
[ ] Engineering drawings reflect unit modifications
[ ] Modifications have received a process safety review
[ ] Operators follow written instructions
[ ] Equipment correct for the process
[ ] Reactants handled and stored correctly
[ ] Personal protective equipment in correct location
[ ] Sufficient ventilation
[ ] Electrical equipment compatible with area processes
[ ] Identify hazardous materials involved in the process
[ ] Location of stored hazardous materials
[ ] Quantities of stored hazardous materials
[ ] Emergency procedures to handle accidental events
[ ] Training to employ emergency procedures and equipment
[ ] Identify potential errors in operation
[ ] Identify potential consequences of errors in operation
[ ] Equipment safeguards operable
[ ] Explosion relief venting adequate
[ ] Explosion relief properly directed
[ ] Provisions for uncontrolled reactions
[ ] Training for new process operators

Chapter 7 - Incidents / Emergency Planning / Audits

- [ ] Periodic training for existing operators
- [ ] Emergency training exercises
- [ ] Chemical fire extinguishing system warranted
- [ ] Explosion suppression system warranted
- [ ] Identify most likely possibilities of accidental releases
- [ ] Controls to eliminate contamination of product
- [ ] Storage of contamination sensitive bulk materials
- [ ] Critical spare parts on hand
- [ ] Emergency dump system warranted
- [ ] Critical maintenance procedures identified

PART II - EPA'S CHEMICAL ACCIDENT RELEASE PREVENTION REGULATIONS

Chapter 8

Background on EPA's Chemical Accident Release Prevention Regulations

8.1 THE NEW EPA RULE

In response to the Accidental Release provisions added to the Clean Air Act Amendments of 1990, section 112(r), the EPA has proposed implementation of Risk Management Programs (RMPs) for chemical accidental release prevention, which are even more comprehensive and extensive in coverage than OSHA's PSM Standard requirements. Proposed regulations were issued in October 1993 (58 Fed. Reg. 54190 (Oct. 20, 1993)) and final regulations are expected to be promulgated by the end of 1994.

The final rule will apply only to stationary sources that manufacture, process, use, store, or otherwise handle regulated substances in quantities that exceed specified thresholds. Companies will have three years to comply with the rule after it becomes final.

The EPA regulations will establish requirements for preparing and implementing RMPs at more than 140,000 facilities, many of which are beyond the scope of regulation under OSHA's PSM Standard. The agency estimates the cost of compliance from $1,700 for a facility in the service industry sector, to $153,000 for a large complex manufacturing site. But industry officials claim that this estimate is low and say that a $10,000-to-$250,000 range is more realistic. (See "Air Pollution: Accidental Chemical Release Proposal Would Require Plans From 140,500 Facilities," 17 Chem. Reg. Rptr. 1307 (Oct. 22, 1993).) Under the proposed regulations, firms that do not comply could face civil penalties of up to $25,000 per day per violation, as well as criminal sanctions (58 Fed. Reg. 54190).

198 OSHA and EPA PSM Requirements

Manufacturers constitute the prime target of the proposed rule, but cold-storage facilities that use ammonia as a refrigerant, public drinking water and waste water treatment systems, chemical wholesalers, propane retailers, makers of small explosives, and utilities also could be covered. These facilities may be covered by the new regulations because they use one or more chemicals found on EPA's list of regulated substances. This list was finalized in January 1994 (59 Fed. Reg. 4478 (Jan. 31, 1994.)) and is reproduced as **Appendix 7**. That list includes 100 substances that were listed based on acute toxicity; 62 flammable gases and highly flammable liquids; and high explosives as a class.

The new EPA regulations are similar to OSHA's PSM regulations. EPA has worked closely with OSHA to minimize conflicting requirements, and ensure consistency between the two sets of regulations. The Prevention Program/Management System outlined in the EPA regulations adopts the language from OSHA's PSM Standard almost verbatim. The EPA prevention program includes: process safety information, process hazards analysis, standard operating procedures, training, maintenance (mechanical integrity), pre-startup review, management of change; safety audits; and accident investigation, which are similar in scope and coverage to comparable provisions of OSHA's PSM rule. However, the EPA goes a step further by requiring that a management system be developed to ensure integration and implementation of the risk management elements on an ongoing basis.

8.2   COMPARISON OF OSHA AND EPA RULES

Although there are similarities between the two rules, an understanding of the differences is critical to compliance. While the OSHA regulations primarily govern potential hazards inside regulated facilities, the EPA regulations are more concerned with potential incidents that may occur outside plant boundaries. A key difference is that the EPA regulations will require submission of extensive documentation to LEPCs, SERCs and a Chemical Safety and Investigation Board. RMPs will be made available to the public.

EPA's chemical list contains 43 substances not listed by OSHA, while the PSM Standard covers other substances that the EPA is prohibited by law from listing. Likewise, the threshold quantities (TQs) are not always uniform; 15 substances on EPA's list have TQs that are lower than those established by OSHA.

8.3 ACCIDENTAL RELEASE SCENARIOS

The EPA regulations would require companies to evaluate a range of accidental releases and determine a worst-case release scenario for each location. To determine a worst-case release scenario, the facility would be required to examine the process for handling each regulated substance and assume that all of the regulated substances in the process were instantaneously released and all mitigation systems failed to minimize the consequences of the release, according to the proposal.

Developing a worst-case scenario will require the use of air dispersion models to analyze the distances releases might migrate off-site and the extent to which they could pose a danger to the public. In addition to worst-case release scenarios, EPA would require facilities to analyze other, potentially significant accidental release scenarios for each process in which the regulated substance is used above the threshold quantity.

While examining these potential release scenarios, facilities could assume that passive mitigation systems -such as containment dikes - functioned properly. Active mitigation systems, such as excess flow valves, fail-safe systems, scrubbers, flares, deluge systems, and water curtains, would be assumed to fail.

Four states - California, Delaware, Nevada, and New Jersey - already require facilities to develop risk management plans. The primary difference in the four state programs compared with the proposed federal rule is the level of detail required and the method of implementation. EPA has indicated that facilities that file risk management plans in Delaware, Nevada, and New Jersey will be in compliance with most elements in the proposed rule. Because the California rules are more general and because different localities have interpreted the state's requirements differently, EPA's position is that it is not possible to determine, except on a case-by-case basis, to what extent a California facility would be in compliance with the agency's proposal. (See "Air Pollution: Accidental Chemical Release Proposal Would Require Plans From 140,500 Facilities," 17 Chem. Reg. Rptr. 1307 (Oct. 22, 1993).)

8.4 IMPETUS BEHIND THE EPA RULE

Public awareness of the potential danger from accidental releases of hazardous chemicals has increased over the years as serious chemical accidents have occurred around the world (e.g., the 1974 explosion in Flixborough, England, and the 1976 release of dioxin in Seveso, Italy). Public concern intensified following the 1984 release of methyl isocyanate in Bhopal, India, that killed more than 2,000 people living near the facility. A subsequent release from a chemical facility in Institute, West Virginia, sent more than 100 people to the hospital and made Americans aware that such incidents can and do happen in the U.S.

In response to this public concern over the hazards that exist, the United States Environmental Protection Agency (EPA) began its Chemical Emergency Preparedness Program (CEPP) in 1985, as part of the Agency's Air Toxics Strategy. CEPP was a voluntary program to encourage state and local authorities to identify hazards in their areas and to plan for chemical emergency response actions. In 1986, Congress enacted many of the elements of CEPP in the Emergency Planning and Community Right-to-Know Act of 1986 (EPCRA), also known as Title III of the Superfund Amendments and Reauthorization Act of 1986 (SARA). SARA Title III requires states to establish state and local emergency planning groups to develop chemical emergency response plans for each community. SARA Title III also requires facilities to provide information on the hazardous chemicals they have on site to the states, local planners, and fire departments, and, through them, the public. This information forms the foundation of both the community emergency response plans and the public-industry dialogue on risks and risk reduction.

SARA Title III did not mandate that facilities establish accident prevention programs. However, Congress acknowledged the importance of accident prevention by requiring EPA, under SARA section 305(b), to conduct a review of emergency systems to monitor, detect, and prevent chemical accidents. The final report to Congress, Review of Emergency Systems (EPA, 1988), stated that "... prevention does not depend on a single piece of equipment or a single technique. Prevention must be part of a comprehensive, integrated system that considers the hazards of the chemicals involved, the hazards of the process, the hazards to the community, and the capabilities of facility personnel. None of the elements should be considered in isolation nor should any single technical solution be considered a complete solution to a particular problem. Each change in a facility, process, or procedure will have multiple effects that must be assessed in the context of the entire operation." The report concluded that the key to a successful process safety management system is the commitment of management (facility and corporate) to safety.

8.5 EPA'S ACCIDENTAL RELEASE INFORMATION PROGRAM

Although SARA Title III did not directly address accident prevention except through section 305(b), EPA recognized that prevention, preparedness, and response form a continuum. In 1986, therefore, EPA established a chemical accident prevention program to collect information on chemical accidents and to work with other groups to increase knowledge of prevention practices, encourage industry to improve safety at facilities, and foster increased awareness of prevention, preparedness, and response at the local level. Under this program, EPA developed its Accidental Release Information Program (ARIP) to collect data on the causes of chemical accidents and the steps facilities take to prevent recurrences. EPA also developed a program for conducting chemical safety audits at facilities to learn more about how facilities develop systems to prevent accidents. Through the audit program, EPA trained its regional staff as well as state officials on how to conduct audits. EPA worked with trade associations, professional organizations, labor, environmental groups, and other Federal agencies to determine how best to reach smaller operations, which the SARA section 305(b) study indicated are less aware of risks than larger facilities. EPA has also been an active participant in international efforts related to chemical accident prevention, particularly through the Organization of Economic Cooperation and Development, which has held five international workshops from 1989 through 1991 to discuss issues related to accident prevention, preparedness, and response, and has developed guidelines for member countries.

In addition to EPA's work in this area, other agencies, states, industries, trade associations, and professional organizations have developed programs related to chemical accident prevention. OSHA promulgated its PSM standard on February 24, 1992 (See 57 Fed. Reg. 6356). Four states - New Jersey, California, Delaware, and Nevada - have regulations requiring facilities to prepare and implement risk management plans. The American Institute of Chemical Engineers (AIChE), through its Center for Chemical Process Safety, has published guidance on the management of chemical process safety, as well as guidelines on topics related to hazard evaluation, vapor cloud dispersion modeling, handling and storage practices, and vapor cloud mitigation. The Chemical Manufacturers' Association (CMA) has adopted its Responsible Care superTM program, with which all CMA members must comply to maintain membership. The American Petroleum Institute has developed a similar program (RP 750) for its members. In 1982, the European Economic Community adopted the Seveso Directive (82/501/EEC, as amended), which requires facilities handling certain chemicals to develop a safety report that is similar to a risk management plan. Congress also recognized the need for a chemical accident prevention program at the federal level and included prevention provisions in the Clean Air Act Amendments of 1990.

8.6 CLEAN AIR ACT AMENDMENTS OF 1990

The Clean Air Act Amendments of 1990, signed into law on November 15, 1990, amend Clean Air Act (CAA) section 112 by adding a new subsection (r), which includes requirements related to chemical accident prevention. The goal of CAA section 112(r) is to prevent accidental releases of regulated substances and other extremely hazardous substances to the air and to minimize the consequences of releases by focusing preventive measures on those chemicals that pose the greatest risk.

Section 112(r) has a number of provisions. It establishes a general duty for facilities (i.e., stationary sources) to identify hazards that may result from releases, to design and maintain a safe facility, and to minimize the consequences of releases when they occur. Section 112(r)(3) requires EPA to promulgate a list of at least 100 substances that are known to cause, or may be reasonably anticipated to cause, death, injury, or serious adverse effects to human health or the environment when released to air. EPA is required to set thresholds for each listed substance. The final rule for the list and thresholds was published on January 31, 1994 (59 Fed. Reg. 4478).

CAA section 112(r)(7) requires EPA to promulgate, by November 15, 1993, "reasonable regulations and appropriate guidance" to provide for the prevention and detection of accidental releases and for responses to such releases. These regulations shall include, as appropriate, provisions concerning the use, operation, repair, and maintenance of equipment to monitor, detect, inspect, and control releases, including training of personnel in the use and maintenance of equipment or in the conduct of periodic inspections. The regulations shall require facilities to prepare and implement risk management plans that shall provide for compliance with regulations for managing risk (the "risk management program") and shall include a hazard assessment, a prevention program, and an emergency response program. The list and thresholds promulgated under CAA section 112(r)(3) will determine which facilities must comply with the accident prevention regulations.

The CAA, as amended, establishes a Chemical Safety and Hazard Investigation Board to investigate or cause to be investigated the causes of chemical accidents and to report its findings to Congress, federal, state, and local authorities, and the public. Under the CAA, EPA is also required to conduct studies related to accidental releases, including research on hazard assessments, hydrogen fluoride, and air dispersion modeling.

Finally, CAA section 112(l) requires EPA to develop guidance for states, especially for the registration of sources (facilities). This CAA section also contains the statutory authority for EPA to approve and delegate federal authority to the states. (For further information on EPA's proposed rule on CAA section 112(l), see 58 Fed. Reg. 29296, May 19, 1993.)

8.7 RISK MANAGEMENT PROGRAMS: CLEAN AIR ACT REQUIREMENTS

CAA section 112(r)(7)(B)(i) requires EPA to adopt "reasonable regulations and appropriate guidance" to provide for the prevention and detection of accidental releases and for response to such releases. As appropriate, the requirements shall address the use, operation, repair, replacement, and maintenance of equipment to monitor, detect, inspect, and control accidental releases, including the training of persons in the use and maintenance of equipment and in the conduct of periodic inspections. The regulations shall include procedures and measures for emergency response after an accidental release. The Act requires that the regulations be promulgated by November 15, 1993.

CAA section 112(r)(7)(B)(ii) states:
"The regulations under this subparagraph shall require the owner or operator of stationary sources at which a regulated substance is present in more than a threshold quantity to prepare and implement a risk management plan to detect and prevent or minimize accidental releases of such substances from the stationary source, and to provide a prompt emergency response to any such releases in order to protect human health and the environment."

The risk management plans must include a hazard assessment that evaluates potential effects of an accidental release of any regulated substance. The hazard assessment must include an estimate of potential release quantities and downwind effects, including potential exposure to populations. The assessment also must include a five-year release history, including the size, concentration, and duration of releases, and must consider worst-case release scenarios. The risk management plan must also document a prevention program including safety precautions, maintenance, monitoring, and employee training measures. The final specified element that must be documented in the risk management plan is an emergency response program that provides specific actions to be taken in response to a release to protect human health and the environment, including informing the public and local agencies, emergency health care, and employee training.

CAA section 112(r)(7)(B)(iii) requires that the risk management plans be registered with EPA. The plans must be submitted to the implementing agency, the Chemical Safety and Hazard Investigation Board, the SERC) and the LEPC. These plans must be made available to the public under CAA section 114(c). EPA must establish a system for auditing the risk management programs. EPA must also ensure that plans are updated periodically.

The proposed rule would require facilities to do three things:

> (1) *Register with EPA three years after publication of the final rule in the Federal Register.* The registration would consist of a written form to be sent to EPA headquarters indicating that the facility is covered by the rule, identifying the regulated substances triggering the registration and the quantity of those substances (in ranges) in a process. If the information on the registration changes (e.g., because new chemicals are added, chemicals are

dropped, or the quantity changes), facilities would be required to submit an amended registration form;

(2) *Develop and implement a risk management program that includes a hazard assessment, prevention program, and emergency response program, and maintain on-site documentation of the implementation.* The hazard assessment would include off-site consequence analyses and a five-year accident history. The prevention program would consist of a process hazard analysis, process safety information, standard operating procedures (SOPs), training, maintenance, pre-startup reviews, management of change, safety audits, accident investigations, and a management system. The emergency response program would require emergency response plans, drills or exercises, and coordination with public emergency response plans; and

(3) *Develop and submit to the Chemical Safety and Hazard Investigation Board, the implementing agency, SERC, and LEPC, a risk management plan (RMP)* that would document the results of the risk management program including a summary of the off-site consequence analysis, a list of major hazards, steps being taken to address those hazards (i.e., a summary of the facility's prevention program), a five-year accident history, a description of the emergency response program, and a description of the management system that ensures the safety of the facility and the implementation of the required elements. This plan will be available to the public.

The risk management program addresses the general requirements of CAA section 112(r)(7)(B)(i) for regulations to provide for accidental release detection and prevention. The risk management plan (RMP), addresses the specific requirements of CAA section 112(r)(7)(B)(ii) for a plan that provides governmental entities and the public with information on the hazards found at facilities and the facilities' plans for addressing the hazards. These hazards would be identified and addressed through implementation of the risk management program elements. Therefore, the RMP would summarize the results of hazard assessments and analyses and the implementation of the risk management program requirements. The submission requirements (registration and the RMP) address the requirements of CAA section 112(r)(7)(B)(iii), as does the requirement for a system to audit RMPs.

In addition to CAA section 112(r)(7)(B), CAA section 112(r)(7)(A) authorizes EPA to promulgate "release prevention, detection, and correction requirements which may include monitoring, record-keeping, reporting, training, vapor recovery, secondary containment, and other design, equipment, work practice, and operational requirements." EPA is investigating whether regulations, other than the proposed rule on risk management programs, are necessary to prevent and detect accidental releases.

8.8 COST ESTIMATES

EPA estimates that approximately 140,425 facilities would be affected by its proposed rule. Of this universe, 87,800 would also be covered by the OSHA rule or an equivalent state standard; the costs estimated for these facilities reflect only the costs for registering and developing the hazard assessment and RMP. The remaining 52,625 facilities will only be covered by EPA's proposed rule; the estimated costs reflect the costs of implementing all risk management program requirements. The total universe of covered facilities includes 22,935 manufacturers (covering all manufacturing sectors except tobacco); 3,360 private utilities (electric and gas utilities, drinking water systems, and treatment works); 33,250 public drinking water and treatment works; 50,000 cold storage facilities; 9,460 wholesalers; 20,000 propane retailers, and 1,240 service industry facilities.

EPA estimates that the costs per facility will vary, for facilities covered solely by the EPA rule, from approximately $1,700 for a facility in the service industry sector, to approximately $153,000 for a large complex manufacturing facility. EPA did not estimate the cost of compliance for a highly complex facility such as a petroleum refinery because all of these facilities are covered by the OSHA standard.

The most costly items in the prevention program for manufacturers include the process hazard analysis, which varies from $6,600 per process for a simple facility to $35,000 per evaluation for a complex facility; training costs, which vary from $2,400 for a small simple facility to $61,000 for a large (150-employee) complex facility; process and equipment information, which may cost a large facility $36,000 per process; and SOPs, which vary from $2,500 for a simple process to $14,000 for a complex process.

Costs for non-manufacturers are estimated to be considerably lower both because their operations frequently do not involve special equipment and because model RMPs and guidance are assumed to be available to them, thus lessening the burden. The cost of conducting the hazard assessment is estimated to vary from $70 per assessment for non-manufacturers to $280 per assessment for large complex manufacturers; the number of assessments likely to be required is estimated to vary from one (for a cold storage facility with only ammonia on-site) to ten for a petrochemical facility. The cost of developing the RMP is estimated to range from $156 to more than $1,000 for a highly complex facility. Registration costs are estimated to vary between $43 and $105, depending on the number of substances on-site. EPA estimates that the initial cost of the proposed rule would be $503 million.

EPA estimated subsequent year total costs for a period of ten years. Costs vary from year to year because certain risk management program elements are not required to be updated yearly. For example, safety audits would be conducted every three years; hazard assessments and process hazard analyses would be updated every five years.

Chapter 9

Risk Management Plans

9.1 PURPOSE OF EPA'S NEW RULE

The purpose of EPA's new rule is to require industry to develop an integrated, holistic approach to managing the risks posed by the presence and use of regulated substances. EPA's rule builds on process safety management elements included in OSHA's standard: process information, process hazard analysis, standard operating procedures, training, pre-startup reviews, mechanical integrity, management of change, accident investigation, safety audits, and emergency response. The implementation of these elements and the development of the Risk Management Plans (RMPs) that will be submitted to governmental authorities will assist the owners and operators of facilities to identify hazards and construct a management system that addresses the hazards in a manner that is most effective for the specific circumstances and complexity of the facility.

EPA's rule, particularly the prevention program, emphasizes the importance of management and management commitment for two reasons. First, without management commitment and an integrated system for managing process safety, it is unlikely that safety will be consistently recognized as a priority. Second, although for some facilities better or different technologies may be the most effective methods of addressing hazards, the technologies, by themselves, cannot ensure safety. Equipment must be maintained and workers trained in its proper uses. Changes in the process or procedures may affect the safe operation of technologies. Only with an integrated management system that continually evaluates the safety of a facility can the hazards posed by regulated substances be managed to minimize the likelihood of accidental releases.

Besides lessening the likelihood and severity of accidents, the implementation of process safety management can help facilities run more efficiently. Companies that have instituted risk management programs report reductions in injuries, lost-time accidents, mechanical breakdowns, maintenance costs, and material losses. Safety improvements will result in lower insurance costs. By preventing accidental releases, companies may minimize environmental damage and necessary cleanup costs.

9.1.1 Applicability of the EPA Rule

The CAA states that facilities covered by the risk management program regulations are those that have more than a threshold quantity of a regulated substance based on the final list promulgated by EPA. EPA estimates that approximately 140,425 facilities would be affected by its new rule. Approximately 87,800 of those facilities would also be covered by OSHA's process safety management standard. The largest sectors covered by the rules would be cold storage facilities (which use ammonia as a refrigerant), public drinking water systems and publicly owned treatment works, manufacturers, and propane retailers. Some wholesalers and service industries would also be covered.

The risk management program rules would affect only those areas at facilities where regulated substances are manufactured, processed, used, stored, or otherwise handled. If a facility uses a regulated substance in quantities above a threshold in only one process (e.g., wastewater treatment or refrigeration), only that process (as well as any unloading, transferring, and storing of the substance) would be covered by the rule. If a single process at a facility includes more than one regulated substance, a single process hazard analysis may cover all regulated substances for that process. EPA realizes that some facilities, such as batch processors (e.g., specialty chemical manufacturers), may have regulated substances on-site for limited periods during the year; for example, a batch processor may use a regulated substance for only one month during the year. In some cases, these facilities may not be able to predict accurately which substances they will be handling. However, the agency believes it is important for any facility that handles a regulated substance to have in place a program to manage risks and ensure safe operations. Because regulated substances would not be covered if they represent less than one percent by weight of a solution, EPA does not expect that the risk management program of publicly owned treatment works would need to cover the substances they receive from facilities for treatment.

9.1.2 Important Definitions

A "significant accidental release" means any accidental release of a regulated substance that has caused or has the potential to cause off-site consequences such as death, injury, or adverse effects to human health or the environment or to cause the public to seek shelter or be evacuated to avoid such consequences.

"Worst-case release" would mean the loss of all of the regulated substance from the process in an accidental release that leads to the worst off-site consequences.

9.1.3 Regulated Substances and Thresholds

The Environmental Protection Agency (EPA) has promulgated its final list of regulated substances and thresholds required under section 112(r) of the Clean Air Act as amended (See 59 Fed. Reg. 4478 (Jan. 31, 1994)). The list is composed of three categories: A list of 77 toxic substances, a list of 63 flammable substances, and explosive substances with a mass explosion hazard as listed by the United States Department of Transportation (DOT). Threshold quantities are established for toxic substances ranging from 500 to 20,000 pounds. For all listed flammable substances the threshold quantity is established at 10,000 pounds. For explosive substances the threshold quantity is established at 5,000 pounds.

The list and threshold quantities will identify facilities subject to the chemical accident prevention regulations promulgated under section 112(r) of the Clean Air Act as amended, and discussed throughout this chapter. EPA also promulgated in this regulation the requirements for the petition process for additions to, or deletions from, the list of regulated substances. EPA is deferring action on a proposed exemption from regulation for listed flammable substances when used solely for facility consumption as fuel. This list of substances and threshold quantities became effective on March 2, 1994.

9.1.4 Risk Management Program Elements

The Clean Air Act mandates that the risk management plan document three elements: a hazard assessment, a prevention program, and an emergency response program. Each of these elements of the risk management program is discussed in Sections 9.2 through 9.4.

9.2 HAZARD ASSESSMENT

The Clean Air Act requires a hazard assessment that includes evaluation of a range of releases including worst-case accidental releases; analyses of potential off-site consequences; and a five-year accident history. To allow EPA's prevention program requirements to parallel OSHA's process safety management standard, EPA proposed separating the off-site consequence analysis and five-year accident history from the formal process hazard analysis requirement. The proposed rule would require a hazard assessment that examines a range of accidental release scenarios, selects a worst-case accidental release scenario, analyzes off-site consequences for selected release scenarios including worst case, and documents a five-year history of significant accidental releases and accidental releases with the potential for off-site consequences.

EPA is proposing that facilities complete a hazard assessment for each regulated substance present above the threshold quantity. Facilities that use the regulated substance above its threshold in several locations or processes would need to evaluate a range of accidental releases and determine a worst-case release scenario for each location. The range of releases should include only those events that could lead to significant releases (i.e., accidental releases that have the potential to cause off-site death, injury, or serious adverse effects to human health or the environment).

9.2.1 Worst-Case Release Scenarios

EPA is proposing to define the worst-case release as the instantaneous loss of all of the regulated substance in a process, with failure of all mitigation systems (active and passive). EPA recognizes that this definition may require facilities to consider release scenarios that are highly unlikely. Such a definition will, however, define for the public the extreme worst-case. The proposed definition will also reduce the burden on regulated facilities; a requirement for analysis of a "credible worst-case" would lead to more analyses and documentation to defend the selected scenario. In addition, if each facility defined its own worst-case, local authorities could find it difficult to compare the results.

The agency recognizes that this approach differs from the approach EPA used in its Technical Guidance for Hazards Analysis for local planners to assess credible worst-case releases for purposes of screening out situations with little or no impact. The credible worst case in the guidance assumed that the entire quantity of a substance was released from the largest vessel or group of interconnected vessels. Gases were assumed to be released in 10 minutes while liquids were assumed to be spilled on the ground or in a diked area and allowed to volatilize. Downwind impacts were assessed using conservative meteorological conditions.

The agency still supports this approach for screening, however, the methodology does not fully account for site-specific conditions that affect the rate of release. For example, gases may be stored in a liquefied state or a liquid may be handled in large quantities at higher than ambient temperatures giving much different release rates. The agency believes that the worst-case analysis should account for site-specific conditions and physical chemical properties.

EPA considered defining worst case as the instantaneous loss of the regulated substance from the largest containment vessel or pipeline on-site. This approach is similar to the Technical Guidance approach. However, because the threshold quantity applies to the quantity in a process and the definition of a process defines the vessels and piping to be considered, the worst case should reflect the accidental release that could occur from catastrophic vessel and piping failures.

9.2.2 Other Accidental Release Scenarios

In addition to the worst-case release scenarios, EPA would require facilities to analyze other more likely significant accidental release scenarios for each process in which the regulated substance is used above the threshold quantity. The proposed rule specifies several possible accident causes that facilities should consider when defining these more likely release scenarios. The list, however, should not be viewed as all inclusive. Each facility should examine its processes to determine the event or sequence of events that may lead to significant accidental releases. When examining these potential release scenarios, facilities would be allowed to assume that passive mitigation systems, such as containment dikes, functioned properly. Active mitigation systems, such as excess flow valves, fail-safe systems, scrubbers, flares, deluge systems, and water curtains, would be assumed to fail. The agency plans to issue guidance on the evaluation of a range of accidental releases and determination of the worst-case scenario.

The proposed rule does not specify the number of other more likely significant accidental release scenarios facilities would be required to analyze. Although this approach provides flexibility, it may create uncertainty about what EPA will consider an adequate number of scenarios. EPA is reviewing whether it should specify a minimum number of scenarios to be analyzed, whether the minimum should vary with the complexity of the facility, and what the minimum(s) should be.

9.2.3 Potential Off-site Consequences

Once the worst-case and more likely significant accidental release scenarios are identified, the facility would be required to analyze the potential off-site consequences associated with these scenarios. The off-site analyses would estimate, using models or other approaches specific to each substance, the possible rate of release, quantity released, and duration of the release, and the distances in any direction that the substance could travel before it dispersed enough to no longer pose a hazard to the public health or environment. Facilities would be required to analyze the releases under average weather conditions for the facility and worst-case weather conditions, which would be defined as a wind speed of 1.5 meters per second and F stability (moderately stable weather conditions). For flammables and explosives, the analyses should consider the distances in all directions that might be affected by pressure waves, fire, or debris. The analyses would also identify all populations that could be affected by such a release, including sensitive populations (e.g., schools, hospitals), and would detail potential environmental damage.

The fate and transport of the regulated substances can be evaluated using air dispersion models. EPA has published guidance on conducting similar analyses in its Technical Guidance for

214 OSHA and EPA PSM Requirements

Hazards Analysis, much of which could be useful in developing the off-site consequence analyses. Computer models to estimate the impacts of vapor cloud explosions also are available. EPA, the Department of Transportation (DOT), and the Federal Emergency Management Agency have developed a model - the Automated Resources for Chemical Hazard Incident Evaluation (ARCHIE) - for vapor cloud explosion evaluation. The World Bank's WHAZAN model also evaluates this type of incident, as do other commercially available models. Simple equations can be used to calculate the impacts of explosions at various distances. EPA plans to develop additional guidance to assist facilities in analyzing off-site impacts.

9.2.4 Use of Generic Methodologies for Assessing Off-site Impacts

Although the worst-case scenario is specifically defined, facilities are likely to use different models and approaches to estimate off-site impacts. In addition, facilities may need to use different models and analytical techniques to account for site-specific conditions in assessing off-site impacts associated with other scenarios. The agency recognizes that facilities will need to have in-house expertise or hire consultants with such expertise to complete these off-site impact analyses. This may pose a significant resource burden on some facilities, and the different approaches and models can make the off-site consequence results more difficult for local emergency planners to use. The agency is working on ways to minimize this burden and make the results useful for local emergency planners. For example, the statute requires the Administrator to issue RMP guidance and model RMPs. The Agency is considering the development of a set of simple, generic tools that would be included in the guidance and that could be used for the assessment of off-site impacts. EPA could develop, for example, a generic methodology for assessing the off-site impacts similar to the methodology included in the Technical Guidance for Hazards Analysis cited above. Using a generic methodology for assessing the off-site impacts would allow a more direct comparison among facilities of potential off-site consequences. At the same time, this approach could reduce the resource burden imposed by the rule on many facilities, particularly smaller businesses by reducing the need for consultants to perform the off-site consequence analysis.

The agency recognizes the limitations associated with simple, generic tools that will need to cover a potentially wide variety of scenarios. It would be difficult to construct a generic methodology which includes assumptions about the characteristics of chemicals, the range of chemical processes (e.g., conditions involving high temperatures and pressures), and other site-specific parameters. As a result, a generic methodology will generally be less sensitive to these conditions (or attributes) and may yield overly conservative or less realistic estimates of off-site impacts. EPA is reviewing this approach and considering input on possible innovative ways to assist facilities in off-site impact analysis that might reduce the burden and provide meaningful, useful results.

9.2.5 Range of Events

Specific information on the worst-case scenario will help public emergency planners and responders recognize the maximum hazard potential surrounding the facility. The Agency recognizes, however, that the worst-case scenario may often be highly unlikely in comparison to other release scenarios with lesser potential consequences. Focusing on the worst-case scenario alone, therefore, could lead public agencies and the public to overestimate the threat posed by a facility. For this reason, EPA believes that facilities must examine a range of events in addition to the worst-case scenario and communicate information on these events to public agencies and the public to provide additional information on the hazards posed by the facility. In addition, EPA does not want facilities to focus solely on the worst-case release because other release scenarios are of concern, are generally far more likely than a worst-case release scenario, and must be addressed in the prevention program. Therefore, EPA is requiring facilities to analyze hazards associated not only with the worst-case scenario, but also with more likely significant releases.

9.2.6 Updating of Off-site Consequences Analyses

EPA would require that facilities update the off-site consequence analyses every five years, with the RMP update, or sooner if changes at the facility or its surroundings might reasonably be expected to make the results inaccurate to a significant degree. For example, a substantial increase or decrease in the quantity of a regulated substance could significantly change the distance a substance could travel before dispersing and posing no hazard. Major changes in housing or land-use patterns, such as the construction of new, large-scale housing developments or commercial areas, could change substantially the population potentially affected.

9.2.7 Five-Year History of Releases

A final element of the hazard assessment specified in the Act is a five-year history of releases of regulated substances. EPA interprets the accident history requirement to cover significant accidental releases and incidents that had the potential for offsite consequences because CAA section 112(r) is directed at preventing such releases. EPA is proposing to require the accident history to document releases that caused or had the potential to cause offsite consequences. As mandated by statute, the history must include the substance and quantity released, the concentration of the substance when released, and the duration of the release. EPA is also proposing that the date of the release, time of the release, and any offsite consequences (e.g., evacuations, injuries, environmental effects) be included. EPA believes that for releases of toxic substances, most of the releases that meet the criteria are already reported to the federal or state governments under CERCLA and SARA Title III. Therefore, development of the five-year history of significant accidental releases would create little additional burden on facilities beyond maintaining records.

9.3 PREVENTION PROGRAM

The Clean Air Act requires that the risk management plan include a prevention program that covers safety precautions and maintenance, monitoring, and employee training measures. Although the Act's requirements for the prevention program are general, a consensus exists among industry, professional organizations, labor, public interest groups, and government on what constitutes a good risk management program. In its Review of Emergency Systems, EPA listed elements of good management programs. The American Institute of Chemical Engineers (AIChE) has published Guidelines for Technical Management of Chemical Process Safety, which includes basically the same elements. Delaware, New Jersey, California, and Nevada have each adopted state risk management program regulations that again cover a similar set of elements. The OSHA chemical process safety management standard covers this same set of elements. Labor and environmental groups recommended similar requirements to Congress and the agencies. Therefore, the prevention program proposed by EPA consists of elements that the federal government and several state agencies, as well as trade associations, professional organizations, labor, and public interest groups believe are necessary in order to have an integrated approach to understanding and managing risks associated with regulated substances at a facility. The elements of this integrated approach are consistent with and fulfill the Clean Air Act requirements.

EPA's proposed prevention program adopts and builds on OSHA's process safety management standard and covers nine procedural areas: Process hazard analysis, process safety information, standard operating procedures (SOPs), training, maintenance, pre-startup review, management of change, safety audits, and accident investigation. The degree of complexity required for compliance for each element will depend on the complexity of the facility. For example, development of process safety information would take far more time and would require greater expertise at a large petrochemical facility than it would at a small drinking water system. As they develop plans for implementing the elements, facility owners or operators would have to consider the complexity of their chemical use, the hazards potentially posed by the chemicals, and potential consequences of an accidental release.

The prevention program elements must be integrated with each other on an ongoing basis. For example, each time a new substance is introduced to a process or new equipment is installed, the process hazard analysis must be reviewed, SOPs updated, training and maintenance programs revised, with new training if needed. An investigation of a near miss or a safety audit may reveal the need for revised operating and maintenance procedures, which will lead to revisions to SOPs, training, and maintenance. The investigation or audit may also indicate a need to review the process hazard analysis. The management system should ensure that a change in any single element leads to a review of other elements to identify any impacts caused by the change.

9.3.1 Development of a Management System

Because it is essential that all of the prevention program elements be integrated into a management system that is implemented on an ongoing basis, EPA is proposing that the owner or operator of the facility designate a single person or position to be responsible for the development and implementation of the overall program. At facilities where individual elements of the program are handled by different people or divisions, the names or positions of the people responsible for each element would also be specified and an organization chart or similar document would be required to define the lines of authority. At a small facility, a single person may be responsible for all elements. At a large company, separate divisions may handle emergency response, training, and maintenance; SOPs may be developed separately for each process area; safety audits may be conducted by corporate officials. In such a situation, it is essential that the involved divisions communicate with each other regularly so that the people in charge of training know when SOPs have been revised and that the emergency response personnel know when changes to processes may affect the hazards in a location. The purpose of the proposed management requirement is to have facility management define a system that integrates the implementation of the elements and assigns responsibility for that implementation.

9.3.2 Process Hazard Analysis

The AIChE's Guidelines for Hazard Evaluation Procedures (AIChE, 1985) defines a hazard evaluation (also known as a process hazard analysis) as a procedure intended "to identify the hazards that exist, the consequences that may occur as a result of the hazards, the likelihood that events may take place that would cause an accident with such a consequence, and the likelihood that safety systems, mitigating systems, and emergency alarms and evacuation plans would function properly and eliminate or reduce the consequences."

A process hazard analysis involves the application of a formal technique, such as a "What-If" or a Hazards and Operability Study (HAZOP). (See **Chapter 4** for descriptions of these techniques). Formal techniques provide a method for a rigorous, step-by-step examination of processes, process equipment and controls, and procedures to identify each point at which a mishap may occur (e.g., a valve failing, a gauge malfunctioning, human error) and examine the possible consequences of that mishap, by itself and in combination with other possible mishaps. The result of a properly conducted process hazard analysis is a list of possible hazards of the process at the facility that could lead to a loss of containment and release of a regulated substance. Process hazard analyses must be conducted by people trained in the techniques and knowledgeable about the process and facility being examined. Such evaluations usually require at least two people, with other experts contributing to the process when necessary; a HAZOP may require a core team of five to seven people. For a simple process, the process hazard analysis may take a day or two; for complex processes, the evaluation may take six weeks to three months.

Although each prevention program requirement is important, EPA considers the process hazard analysis the critical element in developing a risk management program. When EPA analyzed the data collected for the Review of Emergency Systems, it was clear that a substantial number of respondents did not recognize the hazards associated with either the chemicals involved or the processes used. For the most commonly used, high-volume chemicals, such as ammonia and chlorine, a large number of facilities were relatively unaware of the hazards involved. A process hazard analysis would help facilities identify hazards and ways to address them. For example, a 1989 explosion and fire at a facility in Baton Rouge, Louisiana, led to a partial loss of pressure, power, and fire water because the power, steam, and water lines were co-located with the lines carrying flammable gases. The losses complicated and prolonged the process of responding to the release, thereby increasing the damage caused by the release. Similar problems occurred at a facility in Norco, Louisiana, where an explosion led to the loss of all utilities. A thorough and properly done process hazard analysis should identify these types of potential hazards and allow facilities to determine how to mitigate the problems. Process hazard analyses also identify situations where major accidents due to control failure (e.g., pressure gauges, overfill alarms) could be prevented by redundant or backup controls or by frequent maintenance and inspection practices.

Many other elements of a risk management program should flow from, or at least be revised based on, the results of the process hazard analysis. Existing standard operating procedures, training and maintenance programs, and pre-startup reviews may need to be revised to reflect changes in either practices or equipment that derive from the process hazard analysis. The process hazard analysis may help define critical equipment that requires preventive maintenance, inspection, and testing programs. It may also help a facility focus its emergency response programs on the most likely and most serious release scenarios. For many facilities, the process hazard analysis may be necessary to help define the worst-case release scenario that generates the worst off-site consequences. A secondary benefit of the process hazard analysis is that it also can be used to identify pollution prevention opportunities. The same changes in procedures, equipment, controls, or chemicals that may lessen the likelihood of an accidental release often increase the efficiency of operations and result in waste minimization. These changes may reduce costs for facilities by improving the consistency and quality of products and by decreasing the amount of waste that needs to be treated.

9.3.3 Application of PHA Techniques

The proposed rule would require facilities to conduct process hazard analyses after determining a priority order for the analyses based on the degree of hazard posed by the processes covered by the rule; that is, the facility would have to conduct its analyses on the most hazardous processes first, where the degree of hazard is related to potential off-site consequences, operating history of the process, and the age of the process. Facilities would be required to use one or more of six techniques: What-If, Checklist, What-If/Checklist, HAZOP, failure mode and effects analysis, or fault tree analysis. Facilities could also use an equivalent methodology provided the facility could demonstrate that the methodology is equivalent to the listed methods.

The complexity of the process hazard analysis procedure will depend on the complexity of the processes to which it is applied. Any of the listed techniques can be used for simple and complex processes although, for simple processes, the simpler procedures, such as the What If, may be more appropriate. Facilities such as wholesalers who load, unload, store, and sometimes repackage regulated substances would be able to use a simple technique such as a checklist to ensure that the substances are stored and handled properly and that fire suppression systems are appropriate for the substances at the facility. Application of the more complex procedures, such as the HAZOP or fault tree, requires considerable technical expertise and may be more appropriate for complex processes, such as those at petrochemical facilities. In some cases, facilities will want to use several techniques; for example, a facility might start with a What If analysis to identify high hazard areas, then use a HAZOP or fault tree method to examine those areas in greater detail. EPA is planning to develop guidance to help facilities select and use process hazard analysis techniques.

The process hazard analysis would require facilities to conduct a systematic examination of the process and procedures to identify ways in which equipment malfunction, human error, or external events could lead to an accidental release. The evaluation would also review the efficacy of prevention and control measures to prevent accidental releases. The team conducting the process hazard analysis would include at least one person knowledgeable in the technique and one knowledgeable in the process. EPA is considering whether the requirement for a person knowledgeable in the technique should be waived for facilities using checklists and what if questions from a model RMP. The team would be required to submit findings and recommendations to the owner or operator, who then would have to document all actions taken in response to the findings and recommendations, including schedules for implementing changes.

9.3.4 Monitoring and Detection Systems

In response to the CAA's requirement that the prevention program include monitoring, EPA is proposing that the owner or operator investigate and document a plan for (or a rationale for not) installing systems to detect, contain, or mitigate accidental releases if such systems are not already in place. Because accidental releases can be limited or mitigated by the use of detection, secondary containment, and mitigation systems, facilities should consider whether the hazards they have identified could be addressed through such systems. The decision on whether such systems are the best way to address the hazard must, however, rest, in the first instance, with the facility's management. In some cases, monitors and detectors do not exist; mitigation systems may not be technically feasible for certain types of releases. In other cases, steps such as improved procedures and maintenance may

222 OSHA and EPA PSM Requirements

provide a more cost-effective approach to controlling the hazards. The purpose of the requirement is to ensure that facilities consider the available options and find the best method for the facility to address accidental releases.

As required by the CAA, the process hazard analysis must be reviewed and updated periodically. EPA is proposing that the process hazard analysis be reviewed and updated at least every five years, which is the same interval specified in the OSHA process safety management standard.

9.3.5 Compiling Process Safety Information

The process hazard analysis must be based on up-to-date chemical and process information, including information on physical and chemical hazards, process technology (e.g., process chemistry, process parameters), and equipment (e.g., equipment specifications and design, piping and instrumentation drawings). As per OSHA, after the effective date of the rule, facilities would also have to document material and energy balances for new equipment in a process that involve a regulated substance above the threshold quantity to ensure that the equipment is appropriately designed for the process. The material balance is intended only for ensuring the proper design basis for the equipment and is not useful for process inventory accounting or measurement of chemical loss. For example, it is necessary to know the flow rate in mass per unit-time to properly design a heat exchanger; however, this flow rate does not give the mass of the substance consumed or lost in a reaction system.

All required process safety information would apply only to affected equipment, not the facility as a whole. Chemical information is available from Material Safety Data Sheets (MSDSs) mandated under OSHA's Hazard Communication Standard (29 CFR 1910.1200). The level of process technology and equipment information would vary with the type of facility. For warehouses, wholesalers, and service industries, little equipment information would be needed unless special equipment is used with the regulated substances. For manufacturers, more extensive information would be required, including flow charts, piping and instrumentation diagrams of the facility as it currently exists, and electrical, relief, ventilation, and safety system specifications.

9.3.6 Standard Operating Procedures (SOPs)

The results of the process hazard analysis, information developed during the design of a process, and industry and facility experience combine to define the proper way to conduct operations and maintain equipment. SOPs describe the tasks to be performed by the operator, the operating parameters (e.g., temperature, pressure) that must be maintained, and safety precautions needed for both operations and maintenance activities. SOPs must specify the consequences of deviations from safe operating limits (e.g., if the safe operating temperatures are between 100 and 150 degrees Centigrade, the SOPs should indicate what happens if the temperature is above or below those limits). Written SOPs provide a guide to safe operations in a form that can be used by employees. Lack of SOPs and inadequate SOPs have been implicated in a number of catastrophic accidents. For example, improper maintenance procedures have been blamed for a release and explosion at a facility in New Castle, Delaware, in 1980, which killed six people, injured 27 others, and caused more than $63 million in property damage to the facility.

SOPs, which define the proper steps to take in these emergency situations provide a quick source of information that can prevent or mitigate the effects of accidents. SOPs also provide workers and management a standard against which to assess performance; the procedures clarify for both operators and supervisors how operations should be carried out at the facility. Application of SOPs can result in more cost-effective operations by ensuring that operators adhere to procedures that maximize both the safety and efficiency of a process.

EPA is proposing that each facility develop witten SOPs for each process and operation involving the regulated substance above the threshold. The SOPs would include instructions on steps for each operating phase (e.g., initial startup, normal operation, emergency shutdowns, normal shutdowns, emergency operations), operating limits, safety and health considerations, and safety systems. The facility would also be required to provide for control of hazards during operations involving lockout/tagout, confined space entry, and opening process equipment or lines. The facility would also need SOPs to control entrance to the facility by support personnel.

The level of detail included in the SOP should be appropriate for the operation covered. For example, instructions for proper storage of chemicals may be relatively brief, while procedures for routine startup of a complex process may require considerable detail to ensure that each action required is detailed and explained. EPA emphasizes that the SOPs should be usable by the operators in running the process; that is, the SOPs should be written in a language and at a level appropriate for the operators.

9.3.7 Training Requirements

Training provides employees with the information needed to understand what they must do to operate safely and why safe operations are necessary. The required training program is the key to ensuring the effectiveness of other program elements such as SOPs, maintenance programs, pre-startup reviews, and emergency response. Refresher training ensures that employees are reminded of appropriate procedures periodically. Training programs often provide immediate benefits to facilities because trained employees have fewer accidents, damage less equipment through mishandling, and conduct more efficient operations. Inadequately trained maintenance workers have been implicated in the 1989 disaster in Pasadena, Texas, which killed 23 people, injured 130 others, and destroyed $750 million of property at the facility. In 1988, at a plating facility in Auburn, Indiana, untrained workers used hydrochloric acid to clean a tank that had held zinc cyanide. The resulting hydrogen cyanide killed five workers and sent more than ten others to the hospital.

The proposed rule would require each owner or operator to train employees in applicable and appropriate SOPs and provide refresher training at least once every three years. Employers would also be required to ensure that each employee is competent to operate the process safely. EPA is not proposing specific standards for the training requirements because the agency believes that each facility should have the flexibility to develop a training program that reflects its individual situation. Facilities that handle but do not process regulated substances (e.g., many facilities in the non-manufacturing sector) may provide relatively brief training because the procedures to be taught involve a few simple steps. For a complex manufacturing facility, training may take much longer for some operations. For some facilities, formal group training programs may be feasible; for small facilities, one-on-one training may be more appropriate.

Chapter 9 - Risk Management Plans   225

The form of the training program is less important than that relevant training is delivered in a manner most likely to be understood. Facilities would be required to document their training programs to indicate when employees were trained. EPA is also not proposing specific means of ensuring that the training is understood, such as testing, but would simply require that the owner or operator develop a system for ensuring competence and document that system. The proposed rule would require facilities to evaluate the effectiveness of the training and develop a schedule for reviewing and revising the training.

9.3.8   Maintenance (Mechanical Integrity)

The Clean Air Act specifies that the prevention program must include requirements for equipment maintenance. Preventive maintenance, inspection, and testing of equipment are critical to safe operations at a facility. Waiting for equipment to fail often means waiting until an accidental release occurs before addressing a problem. This approach is not acceptable, especially considering the extremely hazardous characteristics of the regulated substances. Preventive maintenance, inspection, and testing are needed because many of the potential failures are not obvious from visual inspections.

For example, failed alarm systems or detectors may need to be tested to determine if they are functioning properly; detectors and monitors, which can provide early warnings of releases, must be calibrated periodically; corrosion of vessels and piping, a hazard with many chemicals, can be detected through testing well before the vessels or pipes fail; scheduled cleaning, oiling, or replacement of parts can prevent equipment failure. A large number of the accidents reported in the Marsh and McLennan review of the 100 largest losses in the petrochemical industry (Large Property Damage Losses in the Hydrocarbon-Chemical Industries, a Thirty-Year Review, 1990) were the result of equipment failure that might have been avoided through preventive maintenance. A 1978 fire and explosion at a Texas City, Texas, facility that led to almost $100 million in property damage was attributed to instrument failure and a faulty relief valve. A 1989 accident in Richmond, California, that injured workers and responders was caused by a failed weld.

Besides preventing accidental releases, maintenance programs also provide direct benefits to facilities by decreasing

226 OSHA and EPA PSM Requirements

the amount of costly down-time that can result from failed equipment. Even in incidents where there is serious property damage, the lost business costs can be significantly greater than the property damage resulting directly from an accident.

EPA is proposing that facilities develop and implement a maintenance program, with written maintenance procedures and training for maintenance workers, for equipment and controls whose failure could lead to a significant accidental release. This equipment may include pressure vessels, storage tanks, piping systems, relief and venting systems, emergency shutdown systems, and controls such as monitors, alarms, sensors, and interlocks. Covered equipment should be inspected, tested, and subject to preventive maintenance. The intervals for such maintenance would depend on the equipment and how it is used. Manufacturers' recommendations may be used to set such schedules and determine testing procedures, but the applicability of those recommendations should be reviewed in light of industry and facility experience and the results of the process hazard analysis. In some cases, facilities will need to schedule more frequent inspections based on their specific uses or experience with equipment failure rates, or because the process hazard analysis indicated that failure of a particular piece of equipment could result in a catastrophic loss of containment. Facilities would be required to replace or repair, in a timely manner, any equipment that is found to be outside acceptable limits.

Facilities would also be required to develop procedures to ensure that replacement equipment and parts meet design specifications. Owners and operators would be required to document their maintenance program, including the written procedures, the schedules used, and the results of each inspection and test performed. The level of complexity and detail in the maintenance program would be directly related to the complexity of the operations and equipment.

9.3.9 Pre-Startup Review

Startup of a new or modified system can be a particularly hazardous time for facilities, especially for complex processes and those that require high temperatures, high pressures, or potentially exothermic reactions. However, even simple facilities need to conduct such reviews. For example, before a chemical distributor accepts a new regulated substance, the distributor should check that the fire suppression system is appropriate for the substance, that workers know how to handle and store the substance, and that emergency response procedures are in place to handle an accidental release.

To help ensure safety during startup, EPA is proposing that all critical systems be checked prior to startup of a new or substantially modified process. A new system would require a process hazard analysis prior to startup. A substantially modified process would include any process where the changes to the process are significant enough to require a reevaluation of the hazards involved because new hazards may have been created as a result of the changes. This review would include a list of items that operators would need to check or test before beginning an operation. Each pre-startup review should ensure that SOPs are in place and training has been conducted.

9.3.10 Management of Change

Chemical processes are integrated systems; changes in one part of the process can have unintended effects in other parts of the system. For example, installation of better seals may increase the pressure in vessels. It is, therefore, important that all changes in processes, chemicals, and procedures be reviewed prior to their implementation to identify any potential hazards that may be created by the modification. Although most changes at facilities are intended to improve safety and efficiency, any modification can have unintended effects and requires a specific review of the safety implications of the change.

Other process modifications are instituted in response to a specific problem that arises unexpectedly. It was such an unexamined change in the installation of a temporary bypass at Flixborough, England, that led to the 1974 release and explosion that killed 28 employees, injured 89 people, and damaged almost 2,000 properties off site. Therefore, EPA is proposing to require management of change procedures. These procedures are important for two reasons: (1) They help facilities evaluate changes and prevent accidents caused by unintended effects from alterations of equipment, procedures, and chemicals; and (2) they ensure that the process safety information and process hazard analyses are kept up-to-date.

Under the proposed rule, the owner or operator of a facility would be required to evaluate every change in equipment (except changes that satisfy the design specifications of the device replaced), processes, chemicals, or procedures to ensure that the technical basis of the change is documented and that the change does not create new hazards; if new hazards are created or if the change results in different procedures being needed, these hazards and changes would need to be addressed prior to implementation. Training, SOPs, and maintenance programs may need to be revised as a result of changes; the process hazard analysis and hazard assessment may need to be revised as well.

9.3.11 Safety Audits

An important tool in ensuring that the process safety management elements are being implemented is the periodic safety audit. The safety audit provides management with a mechanism for oversight of the implementation of the safety elements and of the overall safety of the facility. Safety audits may take many different forms; some facilities use audits to check on compliance with specific regulations, some do spot-checks of safety practices, while others review all key aspects of safety management.

The proposed regulations would require facilities to conduct a complete safety audit once every three years to ensure that the process safety management elements are in place, updated, and being implemented properly. Although compliance with the proposed elements will provide an indication of safe operations, other considerations are important as well. For example, it is not enough to develop and train employees on standard operating procedures; the facility must check to see that procedures are being followed. Therefore, a safety audit is more than a review of regulatory compliance; it is a check, by management, that the facility is being operated safely.

Facilities would be required to document their audits in a report that includes findings and recommendations. Management's response to the findings would also be documented. EPA chose the three-year interval to be consistent with the OSHA requirement for safety audits. EPA notes that for large facilities and those with a number of covered processes, the audit would not need to be performed at one time. The facility may choose to audit different processes on different schedules. The proposed rule would require only that over each three-year period, all covered processes are audited.

9.3.12 Accident Investigation

Accidents can provide valuable information about hazards and the steps needed to prevent accidental releases. Many times, the immediate cause of an accident is the result of a series of other problems that need to be addressed to prevent recurrences. For example, an operator's mistake may be the result of poor training, inappropriate SOPs, or poor design of control systems; equipment failure may result from improper maintenance, misuse of equipment (operating at too high a temperature), or use of incompatible materials. Without a thorough investigation, facilities may miss the opportunity to identify and solve the root problems.

Therefore, EPA is proposing that facilities investigate each significant accidental release. A significant accidental release is one that caused or had the potential to cause off-site death, injury, or serious adverse effects on human health and the environment. EPA notes that significant accidental release does not include near misses. EPA agrees with AIChE that "while it is important to investigate all incidents, as the lessons learned in preventing future incidents are not at all related to the magnitude of the occurrence, it is unquestionable that, at the very least, 'major incidents' should be investigated" (Guidelines for Technical Management of Chemical Process Safety).

EPA encourages facilities to investigate all accidental releases, but would require only that significant accidental releases be investigated. EPA defines significant accidental release as "any accidental release of a regulated substance that has caused or has the potential to cause off-site consequences such as death, injury, or adverse effects to human health or the environment or to cause the public to shelter-in-place or be evacuated to avoid such consequences." EPA is reviewing this approach to define the range of incidents requiring accident investigation. In particular, the agency is considering whether this definition covers too broad or too narrow a set of incidents, and is reviewing comments on any alternative definition that provides greater regulatory certainty.

9.3.13 Evaluation of Root-Causes and Near-Misses

The accident investigation would determine, to the extent possible, the initiating event that led to the release, and the root cause(s). EPA emphasizes that identification of the root causes (e.g., misdesigned piping run) may be more important than identification of the initiating event (e.g., failed flange). The investigation would be summarized in a report to management; the report would include recommendations for steps that need to be taken to prevent recurrences (e.g., piping design review) and improve emergency response and mitigation measures. Management would be required to document its decisions on the recommendations. As with the management of change procedures, the degree of the accident investigation and documentation will vary with the potential seriousness of the accident. For example, a minor release that was prevented from becoming a major release only by prompt action of operators may require more investigation than a large release that can be quickly attributed to single failure (e.g., a faulty high-level alarm).

EPA is also concerned about near misses. Investigation of such incidents may provide facilities with important information on problems that should be addressed before a significant accidental release occurs. Information on near misses could help the agency and facilities understand how accidents occur and how they can be prevented. EPA does not consider a release that occurred, but did not affect the public or the environment because of favorable weather conditions at the time of the release, a near miss. EPA considers this incident a significant accidental release and, therefore, it needs to be investigated.

A near miss would refer to mishaps that did not result in a release for some reason other than explicit system design. For example, a release from a pressure relief valve that is vented to a scrubber would not be a near miss because the system is designed to ensure that relief valve releases are contained and treated. A near miss is a mishap that did not result in a release because of employee actions or luck. For example, a runaway reaction that is brought under control by operators is a near miss and should be investigated to determine why the problem occurred. EPA is considering whether facilities should be required to investigate near misses and on how a near miss should be defined.

9.4 EMERGENCY RESPONSE

The Clean Air Act specifies that the emergency response program include actions to be taken to protect human health and the environment in response to a release, including informing the public and local agencies, emergency health care, and employee training. Emergency response procedures are a necessary part of a risk management program because accidents do happen even with the best safety systems in place. Emergency response procedures can reduce the severity of a release and protect employees, emergency responders, and the public from harmful exposure to the regulated substances. The damage from accidents and risks to responders can be increased if releases have the potential to damage or destroy utilities and equipment needed to respond to the incident. The emergency response plan helps define these worst cases and develop an approach to prevent potential problems.

EPA is proposing that each facility develop an emergency response plan that defines the steps the facility and each employee should take during an accidental release of a regulated substance. The plan would include both evacuation or protective action procedures for employees not directly involved in the response to the release, and the actions taken by employees responsible for responding to and mitigating the release. All employees would be trained in applicable emergency response procedures. The emergency response plan would include descriptions of all response and mitigation systems.

The emergency response plan would also include procedures for notifying the public of releases and of appropriate protective actions and procedures for notifying public agencies. The facility would be required to develop information on proper first-aid and emergency medical care necessary to treat accidental human exposure. EPA is also proposing that the facility emergency response plan be coordinated with the LEPC plans required under EPCRA for chemical releases. Upon request of the LEPC, the facility would be required to provide the LEPC with information necessary to develop and implement the LEPC plan. This requirement is a restatement of the mandate of EPCRA section 303(d)(3) and would be included in this rule to ensure that the facility and community planning efforts are coordinated, which will improve both plans, thereby facilitating effective response actions when releases occur. Facilities would be required to develop written procedures for the use of emergency response equipment and for its maintenance, inspection, and testing. Facilities would be required to conduct drills or exercises to test facility plans and revise the plans based on the results; facilities would be responsible for determining the number and type of drills or exercises they need to conduct and the frequency of these tests.

232 OSHA and EPA PSM Requirements

Most facilities are already required to have at least part of the emergency response plan in place. OSHA requires emergency action plans (29 CFR 1910.38(a)). Facilities that are subject to OSHA's and EPA's Hazardous Waste Operations and Emergency Response (HAZWOPER) rules (29 CFR 1910.120 and 40 CFR Part 311) also must conduct training for their facility response personnel. Facilities covered by EPA's RCRA regulations (40 CFR Parts 264 and 265) or by Spill Prevention Control and Countermeasure rules (40 CFR Part 112) also are required to have many of the emergency response elements in place. EPA is reviewing how the proposed requirements can be best integrated with these existing programs to minimize duplication.

9.5   RISK MANAGEMENT PLAN AND DOCUMENTATION

EPA is proposing that a risk management plan (RMP) be submitted to the implementing agency, Chemical Safety and Hazard Investigation Board, the SERC, and to the LEPC, and be made available to the public. EPA is proposing to make a distinction between the RMP that is submitted to these agencies (and through them to the public) and the documentation supporting the implementation of the risk management program elements that a facility would be required to maintain on site for inspection by EPA and other agencies.

The purpose of the RMP is two-fold: First, to provide government agencies and the public with sufficient information to understand the hazards at the facility and the approach the facility is using to manage the risks and, second, to have the facility develop an ongoing system for managing implementation of safety practices and procedures. The information provided in the RMP will assist government agencies in assessing the quality and thoroughness of a facility's risk management program. Because of the large number of potentially affected facilities, it is unlikely that EPA or the state implementing agency will audit a substantial percentage of the facilities in any one year. Consequently, it is important that government agencies have enough information in the RMP to identify those facilities that pose the greatest potential hazards, either because of the quantity and kind of substances in use or because of prevention practices.

The RMP information also will assist local emergency planners. Under SARA Title III, local planners have received information on substances and quantities at facilities. The RMP will add to these data by providing information on hazards and practices. For example, a large facility with a well-implemented risk management system may pose less of a hazard than smaller facilities, with smaller quantities of chemicals, that have weak programs. With this information, local planners will be better able to focus on facilities that pose the greatest risk and target their work with facilities to improve prevention practices. The public will be able to identify hazards and risk management procedures from the RMP without having important information obscured by detailed submissions.

The second purpose of the RMP is to assist facilities in integrating the risk management program elements. Each facility will approach the management of its hazards in a way that is appropriate for its specific situation. For small facilities, one person may be responsible for implementing and integrating the elements. In large corporations, many of the elements may be handled by different operating divisions. The RMP would include information on the management system the facility uses to integrate the elements and ensure responsibility for the program. EPA thinks that this is an essential step in successful implementation of the program because unless management is accountable for safety and makes it a priority, other employees may not consider safety important. Equally important, by reporting on how it is addressing each of its major hazards, the facility would have to explain how it has applied the various risk management program elements to prevent accidental releases.

9.5.1 RMP Content

The proposed rule would require facilities to submit an RMP that includes the following information:

(1) A copy of the registration form;
(2) A summary of the off-site consequence analyses including worst-case and other more likely release scenarios;
(3) The five-year history of significant accidental releases for each regulated substance;
(4) A list of the major hazards defined through the process hazard analysis, the consequences of failure to control each major hazard, the steps management is taking

or planning to take to address the hazards, and an implementation schedule for each step listed;
(5) A summary of any risk management program elements not covered under the steps taken to address specific hazards (e.g., if training has not been revised to respond to any listed hazard, a summary of the training program would be
needed);
(6) A summary of the facility's emergency response program, including dates and schedules for drills completed and planned, information on coordination with the public, procedures for notifying and alerting the public of a release, and the name of the person responsible for coordinating with public agencies;
(7) A description of the management system used to implement and integrate the elements of the hazard assessment, prevention program, and emergency response program; and
(8) A certification of the accuracy and completeness of the information.

EPA envisions the RMP to be comprehensive and succinct. The off-site consequence analysis information should be a summary of the documentation already developed during the hazard assessment. To keep the size of the RMP manageable, EPA is considering whether it should specify a maximum number of release scenarios a facility may submit as part of its off-site consequence analyses. Complex facilities may conduct a substantial number of such scenarios; submission of every scenario analyzed could overwhelm the user and make the information less useful.

The accident histories can be presented as tables or lists. EPA is not proposing that facilities include every hazard identified through a process hazard analysis, but rather that the RMP include only those hazards that have the potential to lead to significant accidental releases with off-site consequences. For each item included in the RMP, the documentation required by the rule would serve as supporting information.

The information provided should be brief. For example, if corrosion in piping is a hazard, the facility would list corrosion in piping followed by any steps taken to control corrosion and to ensure that corroded pipes are replaced before a release occurs. These steps might include periodic ultrasonic testing, replacement

Chapter 9 - Risk Management Plans 235

of pipes, or something similar. For facilities where the steps taken to address hazards apply to several hazards, the hazards can be grouped under the steps. For example, if revised operating procedures and training were used to control and prevent a number of hazards, the facility could list operating procedures and training followed by the hazards to which they apply. In this way, duplicative entries can be minimized. The length of the list of hazards would vary with the complexity of the facility and with the current state of prevention practices.

EPA is proposing an RMP that summarizes the program because the Agency believes that the information of most use to the public and local agencies will be related to the hazard assessment and consequence analysis, as well as general descriptions of hazards at the facility. Other detailed information is likely to be of little interest and, if submitted, could overwhelm the ability of local agencies to manage and use the information. EPA also believes that the RMPs should not include information that facilities can legitimately claim as confidential business information under CAA section 114(c). The RMP should provide local and state agencies and the public with sufficient information to determine if additional information is needed. The information will be available, if needed, to EPA or state officials conducting audits or compliance inspections. EPA is reviewing the RMP content, particularly the information that communities, local authorities, and public interest groups will find useful in assessing the hazards posed by facilities. EPA is also considering the kinds of information facilities consider confidential (and how facilities can report on hazards without revealing confidential data).

9.5.2 Submission of RMPs

EPA is proposing that the RMP shall be submitted to the Chemical Safety and Hazard Investigation Board, to the implementing agency, the state, and to local emergency planning committees. EPA is concerned about the burden such submissions may place on the entities receiving the RMPs. If each RMP is submitted, the Board could receive more than 140,000 plans; some states could receive several thousand documents. At the local level, the number could vary from a few to more than 50 plans.

EPA is considering three options that might lessen the burden. First, EPA could develop computer software that would provide facilities with standard formats for completing the

236 OSHA and EPA PSM Requirements

information required in the RMP. The RMP could then be submitted on disk in a format that would allow the government agency to locate information quickly. EPA recognizes that while this approach might ease storage problems and related burdens for the Board and the states, many local entities are not equipped to receive documents on disk. In addition, many of the smaller facilities covered by the rule may not yet be computerized. Therefore, this approach would work for only part of the facilities and recipients. The second option would be to allow local authorities to designate the state as the receiving entity, thereby lessening the burden on the local authorities. The third approach would be to require that the RMP be submitted only on request from the Board, state, or local entity. Facilities would be required to develop the RMP and keep a copy available on-site, but would submit it only if requested. EPA is reviewing these approaches and specifically has asked for suggestions on other ways EPA might be able to facilitate the management and use of the RMP information by state and local agencies.

9.5.3 Auditing of RMPs

Section 112(r)(7)(B)(iii) requires EPA to establish, by rule, a system for auditing RMPs and requiring revisions where necessary. EPA is proposing that facilities be selected for audits based on a number of criteria. Specific accidents at a facility or the facility's five-year history of accidents would be one criterion used to select a facility for an audit; similarly, if other facilities in the same industry show a pattern of accidents with regulated substances, a facility might be selected for an audit to ensure that it is addressing the kinds of hazards causing releases at similar facilities. The quantities of regulated substances or the presence of specific regulated substances would also be criteria. For example, facilities with high volumes of one or more regulated substance might be selected, or the audits might focus on particular substances. The location of the facility would be a criterion for selection; facilities close to populated areas, or sensitive populations or ecosystems might be audited because of the potential hazard they pose. The hazards identified in the RMP would be a criterion for selection. Finally, facilities might be randomly selected to provide neutral oversight. EPA is considering whether major facilities should be audited on a regular schedule (e.g., every three to five years).

Chapter 9 - Risk Management Plans   237

The audit is designed to cover the adequacy of the RMP. If, based on the audit, the implementing agency decides that revisions to the RMP are needed, the agency would issue a preliminary determination explaining the basis for the revision and a timetable. This preliminary determination shall include an explanation for the basis for the revisions, reflecting industry standards and guidelines (such as AIChE/CCPS guidelines and ASME and API standards) to the extent that such standards and guidelines are applicable, and shall include a timetable for their implementation.

The owner or operator would have 90 days to respond to the preliminary determination in writing, either agreeing to implement the changes or rejecting the revisions, in whole or in part, with an explanation for any rejection. In its response, the owner or operator may develop substitute revisions addressing the same issues addressed in the preliminary determination. After providing the owner or operator an opportunity to respond, the agency would issue a final determination, which may adopt or modify proposed revisions, or may adopt substitute revisions proposed by the facility. A final determination that rejects a substitute revision would explain the reason for the rejection. Thirty days after the final determination, the facility would be considered to be in violation of the rule unless the RMP is revised. The public would be assured access to preliminary determinations, responses, and final determinations.

In addition to the RMP, the facility would be required to maintain on-site documentation of its process hazard analysis, off-site consequence analysis, process information (e.g., P&IDs, MSDSs), training and maintenance programs, SOPs, pre-startup review list, management of change procedures and records, compliance audits, accident investigation procedures and reports, and emergency response plans. This documentation would include schedules for starting and completing actions based on the recommendations of the process hazard analysis, safety audit, and accident investigation. These documentation requirements are similar to those imposed under OSHA's PSM Standard.

9.5.4 Registration: Information Required

The Clean Air Act requires that RMPs be registered with EPA prior to the effective date of the regulation. EPA is proposing that, within three years of the date of publication of the

238 OSHA and EPA PSM Requirements

final rule, all facilities register with EPA if they have a regulated substance in a quantity that exceeds the threshold quantity. EPA is proposing a simple registration that would require most facilities to complete a one-page form; facilities with large numbers of regulated substances may need an additional page to list the substances. The registration would ask for the name and address of the facility, the facility's Dun and Bradstreet number, the regulated substances on-site, quantities of the substances (in ranges), and the facility's Standard Industrial Classification (SIC) code that applies to the use of each substance. If, at any time after the registration is submitted, the information becomes inaccurate, the facility would be required to file an amended registration within 60 days with the Administrator and the implementing agency.

The association of SIC codes with specific substances would allow EPA to identify the types of processes in which a facility may use the substance without requiring the facility to provide detailed information during registration. The Dun and Bradstreet number is a common identifier for facilities and would allow EPA to cross-reference the data with other EPA databases. Most of the information requested is already reported under SARA Title III. The reporting ranges proposed are the same ranges used for SARA Title III reporting.

EPA is proposing a registration requirement for several reasons. First, the statute requires that RMPs be registered with EPA. Second, EPA is required to establish a system for auditing RMPs. To implement an auditing system, EPA and state agencies that implement the program need to know which facilities are covered by the rule as well as the chemicals they have on-site. Facilities may be selected for auditing based on location, quantities of chemicals on site, specific chemicals, or other criteria. A central source of information on which facilities are covered, for which chemicals, and in which industries is essential to apply criteria for selecting facilities for audits in an equitable manner. Finally, although many of the facilities file similar information with EPA, no current source of data includes all facilities likely to be affected by the proposed rule. EPCRA section 313, for which a national database exists, covers only manufacturers and does not include many of the chemicals proposed for listing. Some of the facilities will be permitted under RCRA, but most will not be. Except for facilities not covered by OSHA's Hazard Communication Standard, most other facilities

Chapter 9 - Risk Management Plans 239

potentially affected by the proposed rule are also covered by EPCRA section 312. However, EPA does not receive section 312 data. Because these data are primarily used at the local level, only a few states have created section 312 databases. In addition, in many states facilities are not required to file chemical-specific information under section 312. Even if every state had a section 312 database, it would not be possible to identify facilities potentially covered by this proposed rule with the section 312 data. Consequently, a separate registration is needed.

EPA considered requiring an earlier registration to help identify potentially affected facilities and disseminate guidance to them. An earlier registration (either 12 months or 24 months after the date of promulgation) would also help states determine the scope of their implementation programs. EPA is considering whether an earlier registration would be beneficial.

9.5.5 Implementation of Registration Requirements

EPA has two main concerns about the implementation of the registration requirement: that multiple or duplicative filings be avoided to the maximum extent possible and that the burden for processing the information be minimized. EPA has requested suggestions on how the registration information might be combined with other forms facilities are required to file to limit the repetitive reporting required of facilities. For example, EPA is considering using the EPCRA section 312 Tier II form as a substitute because the Agency believes this would facilitate integration of CAA activities with SARA Title III activities and would lessen the burden on facilities.

EPA's second concern involves the burden on the government to process the information filed. Each registration will include information that would need to be screened for accuracy. For example, the Chemical Abstract Service (CAS) number and chemical name would need to be checked to make sure that they match and are covered by the rule. SIC codes, Dun and Bradstreet numbers, and quantity range codes would need to be reviewed to ensure that the format (number of digits) and codes were acceptable (i.e., that valid codes were used). Such review could place a substantial burden on EPA and states. EPA is, therefore, considering developing software that would allow electronic filing of the information. The software would perform the quality control function automatically. CAS numbers would be checked to see if they were on the list; the chemical name could then be entered automatically. A list of known synonyms for the listed substances could be included. SIC codes could be checked to ensure that the codes entered actually exist; the format for Dun & Bradstreet numbers could also be reviewed. Messages alerting the facility that the information entered was not acceptable would be provided. Such a computerized form would lessen the time needed to process the information; it would also provide facilities with a quick check on the accuracy of their information and assure them that the data would be accurately represented in EPA's database.

If facilities used such a computerized filing, however, they would still need to submit a signed certification. EPA recognizes that some facilities may not be computerized or may prefer to file a printed form. Although EPA would prefer a computerized filing, printed forms would be acceptable. EPA has requested comments on its plan to encourage computerized filings and specifically solicits suggestions on how such filings could be coordinated with other information filed on disk.

9.6 PENALTIES FOR VIOLATIONS OF THE RULE

Clean Air Act section 112(r)(7)(E) states that after the effective date of the risk management program regulations it shall be unlawful for any person to operate any stationary source subject to the regulations in violation of the requirements of the regulation. Violations of the risk management program and other regulations promulgated under CAA section 112(r)(7) are subject to the same penalties as violations of National Emissions Standards for Hazardous Air Pollutants (NESHAPs) promulgated under CAA section 112(d). Persons in violation of the requirements may be subject to civil penalties of not more than $25,000 per day per violation as well as criminal penalties. Civil penalties may be assessed through court actions or through administrative orders under section 113 of CAA.

9.7 SECTION BY SECTION REVIEW OF EPA'S RULE

EPA is proposing to add a new part 68 to 40 CFR, which would include the risk management program requirements, as well as the list of regulated substances and related regulations, and any additional chemical accident prevention regulations that EPA may

Chapter 9 - Risk Management Plans 241

promulgate in the future. The specific language of each of these proposed provisions is provided in **Appendix 8**.

Proposed Section 68.1 would define the scope of the part.

Proposed Section 68.3 would provide definitions applicable to the Part.

Proposed Section 68.10 would define the applicability of the risk management plan requirements to all stationary sources where a regulated substance is present in a process at any one time in more than the threshold quantity. The section also includes the effective dates for the risk management program elements. Facilities would be required to develop and implement all risk management program elements within three years of the date of promulgation of the rule or within three years of becoming subject to the rule (i.e., three years after the facility introduces a new regulated substance to its operations or a new substance is listed).

Proposed Section 68.12 would define the requirements for registration. Facilities would be required to register three years after the date of promulgation of the rule or within three years of date on which the facility becomes subject to the rule (either because the facility introduces a new regulated substance to its operations or a new substance is listed). If the information submitted on a registration form is no longer accurate, facilities would be required to update the information within 60 days of the change.

Proposed Section 68.15 would provide the requirements for the hazard assessment. Facilities would be required to complete a hazard assessment for each regulated substance present in greater than a threshold quantity. For each such substance, a worst-case release scenario would have to be defined. The off-site consequences of a range of release scenarios, including the worst-case and other more likely significant accidental release scenarios, would have to be analyzed. The proposed section specifies a number of scenarios that should be considered and the information that must be included in the off-site consequence analyses. The section also would require the facility to develop and maintain a five-year history of significant accidental releases and releases with the potential for off-site consequences for each regulated substance. The hazard assessment would have to be reviewed and updated every five years, unless changes necessitated an update sooner. The section would detail the documentation that would be required to be maintained on-site.

Proposed Section 68.20 would explain the purpose of the prevention program and specify that the ten elements of the program must be tailored to suit the degree of hazard present at a facility and the degree of complexity of the operations.

Proposed Section 68.22 would require facilities to designate a person or position responsible for overseeing the development and implementation of the prevention program elements. Where other individuals are responsible for separate elements, an organization chart showing lines of authority would be required.

Proposed Section 68.24 would detail the requirements for the process hazard analysis. A process hazard analysis would be required for each location where regulated substances are present above the threshold quantity. Formal process hazard analysis techniques would have to be applied, with the complexity of the process and potential consequences of a release to be considered in selecting an appropriate technique. The section would require facilities to conduct evaluations on the most hazardous locations first. The process hazard analysis team would be required to report findings and recommendations to management. The facility management would be required to document its response to each finding and recommendation, and maintain a schedule for implementing actions to address findings. If the facility management decides not to implement certain recommendations, a rationale for the decision would have to be documented.

Based on the process hazard analysis results, the facility would be required to evaluate and develop a plan for (or a rationale for not) installing detection and alarm systems, secondary containment and control systems, and mitigation systems. The process hazard analysis would have to be reviewed and updated every five years unless changes of chemical use, process technology, or equipment require an earlier review and revision.

Proposed Section 68.26 would require the facility to develop and maintain up-to-date chemical, technology, and equipment information. Technology information would include process flow diagrams and process chemistry information, maximum intended inventories for vessels, process parameters, and consequences of deviations from parameters. Equipment information would include materials of construction, electrical classifications, material and energy balances, design bases and codes, safety equipment designs, and diagrams of piping, equipment, and controls. The owner or operator would have to document that equipment complies with good engineering practices.

Proposed Section 68.28 would require facilities to develop and maintain written procedures for operations.

Proposed Section 68.30 would require facilities to develop and implement training programs to ensure that all employees are trained in SOPs that apply to them. Refresher training would be required at least every three years. The facility would have to develop a method of ensuring that each employee is competent. In addition, facilities would be required to evaluate the effectiveness of their training. Based on this evaluation, the facility would be required to develop and maintain a schedule for revising the training program. All training conducted at the facility would be documented. In lieu of initial training, the facility could certify that current employees have the knowledge and skills to carry out the SOPs.

Proposed Section 68.32 would require facilities to develop a list of equipment and controls whose failure could lead to a significant accidental release of a regulated substance. For items on the list, a maintenance program that included a schedule for inspections, testing, and maintenance would be required. Inspection and testing procedures and schedules would be based on manufacturers' recommendations unless industry or facility experience indicated that more frequent inspections and tests, or different procedures were needed. Written maintenance procedures and training of maintenance workers would also be required. Equipment found to be outside acceptable limits would have to be replaced or repaired prior to being used again or in a timely manner that ensures safety. Procedures to ensure that replacement equipment is installed properly and consistent with design

244 OSHA and EPA PSM Requirements

specifications would be required. Records of each inspection, test, repair, and replacement would be required.

Proposed Section 68.34 would require facilities to develop procedures to ensure that a pre-startup review is conducted before a new or modified process is brought online. This section would not apply to routine startups after shutdowns for maintenance provided standard procedures are developed for such startups. The pre-startup review would confirm that all installations and changes meet design specifications, that SOPs and maintenance programs are in place for the new processes, and that employees have been trained. Records of each startup, including actions taken to address any problems uncovered during the review, would be maintained at the facility under Section 68.55.

Proposed Section 68.36 would require facilities to develop management of change procedures to ensure that any alteration of chemicals, processes, and procedures are reviewed prior to implementation. Replacement of equipment or controls with a device that meets the design specifications of the replaced device would not be considered a change. The procedures would ensure that the technical basis of the change is documented and that the consequences of the change are evaluated. Process safety information and the process hazard analysis would be updated as needed, as would SOPs, training, and maintenance programs. The results of each such review would be maintained at the facility under Section 68.55.

Proposed Section 68.38 would require facilities to conduct safety audits every three years. Each audit would be documented in a report with findings and recommendations. Management's response to each finding and recommendation would be documented, with a schedule for implementation or a rationale for not implementing.

Proposed Section 68.40 would require facilities to develop and implement procedures to investigate each significant accidental release. Investigations would have to start within 48 hours of the accident. The investigation would document, in a report to management, the initiating event, root causes, and recommendations for preventing recurrences. Management would be required to document its response to each recommendation,

Chapter 9 - Risk Management Plans 245

with either a schedule for implementation or a rationale for not implementing the recommendation. The results of the investigation would have to be reviewed with all potentially affected employees.

Proposed Section 68.45 would require facilities to develop a written emergency response plan that would specify procedures for employees not involved in a response action, procedures for responders, a list of all response and mitigation technologies. The plan would also include procedures for notifying and alerting the public and public response agencies. The facility would be required to have procedures for the use, inspection, testing, and maintenance of response equipment. The facility would also develop information on first aid and emergency health care related to potential exposures. Employees would be trained in applicable response procedures. Facilities would be required to conduct drills or exercises to test the plan. Any drill or exercise would be documented, with findings relevant to plan revisions; management would be required to document responses to the findings, with schedules for implementation. The emergency response plan would be coordinated with the local emergency planning committee's community plan prepared under SARA Title III.

Proposed Section 68.50 would require submission of the RMP containing a copy of the facility's registration form, hazard assessments for each regulated substance (i.e., worst-case scenario, off-site consequences for a range of more likely significant accidental release scenarios, and five-year history of significant accidental releases), a list of major hazards identified through the process hazard analysis, the consequences of failure to control each major hazard, steps being taken to address the hazards, implementation schedules, a summary of other prevention elements, a description of the emergency response plan, a description of the management system for implementing and integrating the risk management program, and a certification of accuracy and completeness. The RMP would be revised and resubmitted every five years unless changes dictate a more frequent revision.

Proposed Section 68.55 would specify which records would need to be maintained and that records would be maintained for five years. Facilities would also be required to maintain

246 OSHA and EPA PSM Requirements

implementation schedules for recommendations from the process hazard analysis, safety audit, and accident investigation.

Proposed Section 68.60 would specify the audit system for reviewing RMPs.

Chapter 10

Comparison Between EPA's CARP Regulations and OSHA's PSM Standard

10.1 RELATIONSHIP BETWEEN EPA'S RULE AND OSHA'S PSM STANDARD

The Clean Air Act Amendments of 1990 (CAAA) section 304 requires OSHA to promulgate a chemical process safety standard and a list of highly hazardous chemicals. To meet this mandate, OSHA promulgated its process safety management standard, which is discussed in detail in Chapters 1 through 7.

EPA and OSHA have met regularly to coordinate their rules to minimize conflicting requirements. To minimize confusion for facilities covered by both rules, the elements and language of EPA's proposed prevention program are, to the maximum extent possible, identical to the parallel elements in OSHA's process safety management standard. The main differences between the EPA's proposed rule and OSHA's standard are those mandated by the CAA, such as the hazard assessment (off-site consequence analysis, the five-year accident history), the emergency response requirements, registration, and the RMP submission to the Board, implementing agency, SERC, and LEPC. In addition, for some elements of the two programs, OSHA's focus is on workplace impacts while EPA's focus is on off-site consequences, reflecting the differing statutory mandates of the two programs. The OSHA PSM Standard includes elements specific to worker issues that EPA has not included in its proposed rule. EPA anticipates that facilities in compliance with the requirements in the OSHA rule also will be in compliance with EPA's proposed prevention program elements. That is, for most prevention program elements, facilities that are in compliance with OSHA's process safety management standard will not need to do anything different or create different on-site documentation to comply with EPA's proposed prevention program requirements.

Because EPA's proposed list of chemicals and thresholds and OSHA's list and thresholds are not identical (EPA covers more substances with acute toxic effects, fewer flammables and explosives, and no reactives) and because OSHA does not cover state and local government employees, the universes of facilities covered by the two rules are not identical, although they substantially overlap.

10.2 DIFFERENCES BETWEEN EPA'S RULE AND OSHA'S PSM STANDARD

The primary differences between EPA's proposed rule and OSHA's process safety management standard are the result of the different statutory requirements for the two rules. The CAA requires EPA to include several elements in its regulation that are not mandated for OSHA. Specifically, EPA's rule must include a hazard assessment, an emergency response program with certain elements, registration, and the submittal and auditing of the RMP. The only other element EPA is proposing that is not included in the OSHA standard is the requirement for the owner or operator of a facility to define its management system and name the person or position responsible for the program. EPA considers the management requirement critical to ensuring that the risk management program elements are integrated with each other on an ongoing basis. EPA expects that this requirement will create no additional burden for facilities because the proposed section would only require facilities to provide the name or names of people or positions responsible for implementing the program.

EPA's proposed hazard assessment includes an off-site consequence analysis and a five-year accident history, as required by the CAA. Under the OSHA standard, facilities are required to develop an on-site consequence analysis. Most of the information needed to define accidental release scenarios will be derived from the process hazard analysis, which would be the same under the two rules. The main differences under the EPA rule would be the need to use air dispersion models to analyze the distances releases might migrate and the need to document the areas potentially affected by the releases. EPA's hazard assessment also is required to include a five-year release history, which would overlap to

Chapter 10 - Comparison of EPA and OSHA Rules

some degree with a requirement in OSHA's process hazard analysis.

EPA's proposed emergency response provisions respond to the language in the CAA and are somewhat different from the OSHA requirement. Under the OSHA standard, facilities must comply with one of two existing OSHA standards. Facilities that are currently in compliance with OSHA's Hazardous Waste Operations and Emergency Response Standard (29 CFR 1910.120) are likely to be in substantial compliance with EPA's proposed rule. OSHA's emergency action plan regulation (29 CFR 1910.38(a)) basically requires an evacuation plan.

The CAA requires EPA's emergency response program to include "specific actions to be taken in response to an accidental release of a regulated substance so as to protect human health and the environment" (CAA section 112(r)(7)(B)(ii)). Therefore, facilities that currently have only an emergency action plan required under 29 CFR 1910.138(a) would, under EPA's proposed rule, need to develop a more extensive emergency response plan that details how the facility would respond to a release to limit off-site consequences.

EPA is also proposing that facilities conduct drills and exercises to test their plans. Without such exercises, a facility will not be certain that a plan can be implemented properly during an emergency. All facilities covered by the EPA rule would need to coordinate their plans with the LEPC, which is not required by the OSHA standard. EPA considers this coordination essential to protect the public. Many facilities are already coordinating their plans with the LEPC plans and with local emergency responders. Therefore, EPA does not anticipate that this requirement will add substantially to the burden for most facilities.

The final differences between the two rules are the proposed requirements for registration, submission, and auditing of the RMP. CAA section 112(r)(7)(B)(iii) mandates these requirements. The information in the RMP would be derived from the documentation required elsewhere under the EPA proposed rule or OSHA's standard. Consequently, EPA expects that the RMP will not add substantially to the burden of complying with the rules.

10.3 SECTION-BY-SECTION COMPARISON OF EPA'S CARP REGULATIONS AND OSHA'S PSM STANDARD

Except for the management system requirement, the proposed EPA prevention program covers the same elements as OSHA's process safety management standard and generally uses identical language except where the statutory mandates of the two agencies dictate differences. EPA has added introductory paragraphs to most sections to provide further information to the regulated community; these paragraphs impose no additional requirements and are intended to clarify the purpose of the section's requirements and the level of detail expected of different types of facilities. In addition, EPA has made editorial changes in the OSHA language to make the rule consistent with the CAA's statutory language. Specifically, where OSHA uses the word "employer," EPA would use "owner or operator," which is defined in the CAA. Where OSHA uses "highly hazardous chemicals," EPA would use "regulated substance." Where OSHA uses "facility," EPA would use "stationary source." Where OSHA uses "standard," EPA would use "rule." Finally, where OSHA references workplace impacts, EPA would reference off-site consequences, reflecting the different statutory mandates of the two agencies.

The specific parallel elements of the two rules are as follows:

EPA's process hazard analysis requirement (Section 68.24) is the same as OSHA's process hazard analysis requirements (29 CFR 1910.119(e)), with the following changes:

(1) An introductory paragraph;
(2) the priority order for conducting the analysis would consider off-site consequences rather than the number of potentially affected employees;
(3) OSHA's schedule for implementation would not be included because the CAA requires that facilities comply with EPA's rule within three years of the date of promulgation and, therefore, OSHA's five-year schedule could not be used;
(4) the identification of previous incidents would be limited to those with off-site consequences rather

Chapter 10 - Comparison of EPA and OSHA Rules 251

than those with catastrophic consequences in the workplace; and

(5) the qualitative evaluation of safety and health impacts would focus on impacts on public health and the environment rather than on employees.

EPA expects that, in most cases, fewer incidents will need to be considered under EPA's proposed rule because releases are generally more likely to affect workers rather than the public. However, some types of releases, such as the release at Bhopal, have their primary impact off site. EPA's rule would ensure that these potential releases are evaluated. Finally, in response to the statutory requirement that the prevention program include monitoring, EPA would add a paragraph (j) requiring facilities to evaluate monitors, detectors, containment or control devices, and mitigation systems.

EPA's proposed process safety information (Section 68.26) is identical to OSHA's process safety information system (29 CFR 1910.119(d)) except for editorial changes and the requirement, in paragraph (c)(5), that the evaluation of the consequences of process deviations include those affecting public health and the environment rather than workers.

EPA's standard operating procedures requirement (Section 68.28) is identical to OSHA's operating procedures (29 CFR 1910.119(f)) except for the introductory paragraph and editorial changes.

EPA's training section (Section 68.30) is identical to OSHA's training section (29 CFR 1910.119(g)), except for the introductory paragraph, editorial changes, and a requirement that facilities evaluate the effectiveness of their training programs and revise the programs, if necessary, based on the evaluation.

EPA's maintenance requirements (Section 68.32) uses the same language as OSHA's mechanical integrity paragraph (29 CFR 1910.119(j)) with certain exceptions. EPA would use the term "maintenance" rather than "mechanical integrity" to parallel its statutory language. EPA would add an introductory paragraph and make editorial changes. In Section 68.32(b), EPA would require the facility to develop a list of equipment that requires

252 OSHA and EPA PSM Requirements

maintenance; the OSHA standard provides a list of equipment. EPA's paragraph (b) includes the OSHA list, but EPA is concerned that for some facilities the list may be too extensive and for others it may not be comprehensive. For example, for warehouses, the only equipment that may need maintenance may be the sprinkler system and the forklifts, neither of which are on the list.

EPA believes the responsibility should be on the facility to develop a list, based on specific facility concerns. EPA would also add an opening paragraph to the OSHA paragraph on inspections and testing and include the word "maintenance" before inspection and testing throughout the paragraph. The inclusion of the word "maintenance" would clarify that equipment should be maintained on a regular basis; for some equipment simple routine maintenance, such as cleaning and oiling, may be all that is necessary; other equipment, such as seals, may be replaced on a regular schedule. EPA's revision would clarify that such maintenance is included in the inspection and testing requirement. EPA would also add language to clarify that training of maintenance workers would be documented in the same manner as other training.

EPA's pre-startup review requirement (Section 68.34) is identical to OSHA's pre-startup review paragraph (29 CFR 1910.199(i)) except for editorial changes, the introductory paragraph, and the requirement in Section 68.34(c)(4) that maintenance as well as operating employees are trained prior to startup and that all employees are trained on any new emergency response procedures. EPA believes these additions are necessary to ensure the safety of the facility.

EPA's management of change requirements (Section 68.36) are identical to OSHA's paragraph (29 CFR 1910.119(l)), except for the introductory paragraph, editorial changes, and a new paragraph (b) in which EPA defines alterations that do not constitute a change. Section 68.36(b) is intended to clarify what constitutes a replacement in kind. EPA would also change paragraph (d)(2) to replace OSHA's "impact of change on health and safety" to "impact of change on likelihood of a significant accidental release."

Chapter 10 - Comparison of EPA and OSHA Rules

EPA's safety audit requirement (Section 68.38) is identical to OSHA's compliance audit paragraph (29 CFR 1910.119(o)), except for the introductory paragraph and editorial changes.

EPA's accident investigation requirements (Section 68.40) are identical to OSHA's incident investigation paragraph (29 CFR 1910.119(m)), except for:

(1) The introductory paragraph and editorial changes to substitute the phrase "significant accidental release" for the word "incident";

(2) the addition, in paragraph (b), of a requirement that the procedures be written;

(3) the requirement in paragraph (c) that incidents that require investigation are those that caused or could have caused off-site consequences rather than catastrophic releases in the work place; and

(4) the addition, in paragraph (f)(4), that the facility identify root causes as well as initiating events.

The OSHA standard includes several requirements that are not covered by EPA's proposed rule, e.g., worker consultation, hot work permits, contractor rules, and trade secrets. EPA believes that worker consultation and hot work permits are worker protection issues and are, therefore, properly in OSHA's area of concern. EPA's trade secret rules for the CAA already are covered in 40 CFR part 2.

10.3.1 Use of Contractors

Finally, although EPA recognizes the importance of contractor competence on safety, EPA believes this issue is primarily one that OSHA should address, as it has in its section on contractors. In addition, EPA believes that contractors are mainly an issue at larger companies, most of which are covered by the OSHA standard. EPA is reviewing whether it should adopt OSHA's contractor paragraph as part of the risk management program requirements.

10.3.2 Compliance Dates

As specified in CAA section 112(r)(7)(B)(i), EPA's rule would become effective three years after the date of promulgation. OSHA's rule will allow facilities up to five years to conduct process hazard analyses. Because the OSHA standard was promulgated prior to EPA's rule, however, EPA does not anticipate that the actual compliance dates for the two rules will differ significantly.

10.4 ROLE OF STATE LAWS

Four states - California, New Jersey, Delaware, and Nevada - have implemented state laws that require certain facilities to develop risk management programs. Although the existing state programs differ in some respects, they address the same basic elements that EPA is proposing in this new rule, except that the California program does not specify a management of change procedure. The New Jersey Toxic Catastrophe Prevention Act (TCPA) program is the most detailed program, specifying to a considerable degree the information required to be developed and submitted; New Jersey also requires that workers pass competency tests after training. The Delaware program provides facilities with more flexibility by specifying less detailed requirements. The California program is the most general of the programs; the California risk management plan program developed by each affected facility is driven by the results of the process hazard analysis, rather than responding to a set of specific mandated requirements.

The primary differences in the state programs relate to their implementation and the chemicals covered. New Jersey, Delaware, and Nevada have implemented their programs at the state level. California has delegated implementation authority to more than 100 administering agencies, which are usually the fire or health departments. New Jersey, California, and Nevada require facilities to submit their plans to the administering agencies for review and approval. Delaware requires facilities to maintain the plan and documentation on-site for state inspectors. California also allows the administering agencies to exempt facilities that meet the thresholds if the agency determines that the facility does not pose a significant risk to the community.

Chapter 10 - Comparison of EPA and OSHA Rules

Each of the states has a different list of chemicals and thresholds. New Jersey's list covers 109 acutely toxic substances; Delaware covers 90 toxic substances, as well as flammables and explosives; California covers all 360 of the EPCRA section 302 list of extremely hazardous substances; Nevada adopted OSHA's list of highly hazardous chemicals. California uses EPA's threshold planning quantities (TPQs) as thresholds for notification and allows local agencies to decide whether a facility must comply; New Jersey and Delaware developed separate and different methodologies for calculating thresholds; Nevada adopted OSHA's thresholds. None of the state lists is entirely consistent with EPA's proposed list.

EPA anticipates that facilities currently in compliance with the New Jersey, Delaware, and Nevada regulations will be in compliance with most elements of today's proposed rule. Because the California rules are more general and because different administering agencies have interpreted the requirements differently, it is not possible to determine, except on a case-by-case basis, to what extent a California facility will be in compliance with EPA's rule.

The Clean Air Act section 112(l) allows EPA to delegate the implementation of the risk management program to states that have an approved program. The criteria for state programs are listed in CAA section 112(l)(5). The Act allows states to adopt the federal program or implement a program that is more stringent. Consequently, the existing state programs will require some revisions to meet EPA's requirements or set more stringent requirements than those established by the EPA rule. EPA expects that most of the needed changes will involve the listing of chemicals and adjusting of thresholds. Other states that are developing state programs to implement these regulations should determine whether they have sufficient statutory authority under their air or emergency planning/community right-to-know SARA Title III programs to adopt the requirements of these regulations. EPA will provide additional guidance for states before the final rule is promulgated.

10.5 OTHER APPROACHES CONSIDERED

The CAA requires facilities that have a regulated substance in quantities greater than the threshold to develop and submit RMPs. EPA recognizes that, for small facilities, even the less

256 OSHA and EPA PSM Requirements

complex risk management program that would be needed for simple processes could create a substantial burden. EPA considered three approaches, therefore, that might reduce this burden. Each of these approaches would create two tiers of risk management programs, a minimal program and an expanded risk management program. The approaches differ on how facilities would be divided between the two tiers.

The first approach considered would be to develop criteria for determining when facilities needed an expanded risk management program. The criteria could be as simple as a multiple of threshold quantity (e.g., an expanded risk management program would be required at ten times the threshold quantity), or would combine the quantity on-site with other factors such as distance to the fenceline, proximity of sensitive populations (e.g., hospitals, schools, residences), similar to the approach used in Delaware.

EPA decided not to propose this approach for several reasons. Facility operators in Delaware and state officials report that this approach is difficult to implement because considerable technical expertise is needed and many smaller facilities and non-manufacturers do not have the expertise in house. In addition, developing a set of criteria that would be appropriate in all situations may not be possible because too many factors influence the hazard posed by a particular process and substance. Using the simple multiple of the threshold quantity would ignore the dangers posed by relatively small quantities of regulated substances in specific circumstances.

The second approach considered would be to have facilities determine whether they needed an expanded risk management program based on the off-site consequence analysis: If the worst-case release could not expose the public or the environment to significant risks, the facility would not need an expanded risk management program. Although this approach is a better way to determine whether the potential risks of a facility merit an expanded risk management program, it is fraught with problems. This approach would create considerable potential for debate and legal disputes over the assumptions facilities use to determine off-site consequences. Assumptions appropriate for one facility or area may not be appropriate for others.

EPA believes that this approach would leave facilities uncertain of the legal status of their decisions and create difficulties for enforcement by governments and citizens. In

Chapter 10 - Comparison of EPA and OSHA Rules 257

addition, given the experience of Delaware facilities, it is likely that many smaller facilities and those outside the manufacturing sector would have substantial difficulty understanding and implementing this approach. EPA notes that most of the facilities potentially affected by the proposed rule are non-manufacturers; less than five percent of the potentially affected facilities are chemical manufacturers or petroleum refineries.

The final approach considered would be to follow the California model and let local or state agencies decide which facilities pose the greatest threat and, therefore, require an expanded risk management program. EPA believes that local agencies are in the best position to identify and evaluate local hazards. However, the viability of this approach rests on the ability and willingness of state or local groups to make these decisions. This approach would impose a considerable burden on state and local authorities. It could also lead to the uneven imposition of requirements on facilities if states or localities chose to cover facilities differently. Some facilities already covered by risk management program rules believe that they have been placed at a substantial competitive disadvantage because they are complying with the state law, while similar facilities in other states are not. An uneven implementation also leaves the protection of the public uneven.

APPENDIXES

Appendix 1

PSM Standard (29 CFR 1910.119)

29 CFR 1910.119. Process safety management of highly hazardous chemicals.

Purpose. This section contains requirements for preventing or minimizing the consequences of catastrophic releases of toxic, reactive, flammable, or explosive chemicals. These releases may result in toxic, fire or explosion hazards.

(a) Application.
(1) This section applies to the following:
(i) A process which involves a chemical at or above the specified threshold quantities listed in Appendix A to this section;
(ii) A process which involves a flammable liquid or gas (as defined in 1910.1200(c) of this part) on site in one location, in a quantity of 10,000 pounds (4535.9 kg) or more except for:
(A) Hydrocarbon fuels used solely for workplace consumption as a fuel (e.g., propane used for comfort heating, gasoline for vehicle refueling), if such fuels are not a part of a process containing another highly hazardous chemical covered by this standard;
(B) Flammable liquids stored in atmospheric tanks or transferred which are kept below their normal boiling point without benefit of chilling or refrigeration.
(2) This section does not apply to:
(i) Retail facilities;
(ii) Oil or gas well drilling or servicing operations; or,
(iii) Normally unoccupied remote facilities.

262 OSHA and EPA PSM Requirements

(b) **Definitions.** *Atmospheric tank* means a storage tank which has been designed to operate at pressures from atmospheric through 0.5 p.s.i.g. (pounds per square inch gauge, 3.45 kPa).

Boiling point means the boiling point of a liquid at a pressure of 14.7 pounds per square inch absolute (p.s.i.a.) (760 mm.). For the purposes of this section, where an accurate boiling point is unavailable for the material in question, or for mixtures which do not have a constant boiling point, the 10 percent point of a distillation performed in accordance with the Standard Method of Test for Distillation of Petroleum Products, ASTM D-86-62, may be used as the boiling point of the liquid.

Catastrophic release means a major uncontrolled emission, fire, or explosion, involving one or more highly hazardous chemicals, that presents serious danger to employees in the workplace.

Facility means the buildings, containers or equipment which contain a process. Highly hazardous chemical means a substance possessing toxic, reactive, flammable, or explosive properties and specified by paragraph (a)(1) of this section.

Hot work means work involving electric or gas welding, cutting, brazing, or similar flame or spark-producing operations.

Normally unoccupied remote facility means a facility which is operated, maintained or serviced by employees who visit the facility only periodically to check its operation and to perform necessary operating or maintenance tasks. No employees are permanently stationed at the facility. Facilities meeting this definition are not contiguous with, and must be geographically remote from all other buildings, processes or persons.

Process means any activity involving a highly hazardous chemical including any use, storage, manufacturing, handling, or the on-site movement of such chemicals, or combination of these activities. For purposes of this definition, any group of vessels which are interconnected and separate vessels which are located such that a highly hazardous chemical could be involved in a potential release shall be considered a single process.

Replacement in kind means a replacement which satisfies the design specification.

Trade secret means any confidential formula, pattern, process, device, information or compilation of information that is used in an employer's business, and that gives the employer an opportunity to obtain an advantage over competitors who do not

Appendix 1 - PSM Standard (29 CFR 1910.119)

know or use it. Appendix D contained in s 1910.1200 sets out the criteria to be used in evaluating trade secrets.

(c) Employee participation.
   (1) Employers shall develop a written plan of action regarding the implementation of the employee participation required by this paragraph.
   (2) Employers shall consult with employees and their representatives on the conduct and development of process hazards analyses and on the development of the other elements of process safety management in this standard.
   (3) Employers shall provide to employees and their representatives access to process hazard analyses and to all other information required to be developed under this standard.

(d) Process safety information. In accordance with the schedule set forth in paragraph (e)(1) of this section, the employer shall complete a compilation of written process safety information before conducting any process hazard analysis required by the standard. The compilation of written process safety information is to enable the employer and the employees involved in operating the process to identify and understand the hazards posed by those processes involving highly hazardous chemicals. This process safety information shall include information pertaining to the hazards of the highly hazardous chemicals used or produced by the process, information pertaining to the technology of the process, and information pertaining to the equipment in the process.
   (1) Information pertaining to the hazards of the highly hazardous chemicals in the process. This information shall consist of at least the following:
      (i) Toxicity information;
      (ii) Permissible exposure limits;
      (iii) Physical data;
      (iv) Reactivity data:
      (v) Corrosivity data;
      (vi) Thermal and chemical stability data; and
      (vii) Hazardous effects of inadvertent mixing of different materials that could foreseeably occur.

264 OSHA and EPA PSM Requirements

<u>Note</u>: Material Safety Data Sheets meeting the requirements of 29 CFR 1910.1200(g) may be used to comply with this requirement to the extent they contain the information required by this subparagraph.

(2) Information pertaining to the technology of the process.
(i) Information concerning the technology of the process shall include at least the following:
(A) A block flow diagram or simplified process flow diagram (see Appendix B to this section);
(B) Process chemistry;
(C) Maximum intended inventory;
(D) Safe upper and lower limits for such items as temperatures, pressures, flows or compositions; and,
(E) An evaluation of the consequences of deviations, including those affecting the safety and health of employees.
(ii) Where the original technical information no longer exists, such information may be developed in conjunction with the process hazard analysis in sufficient detail to support the analysis.
(3) Information pertaining to the equipment in the process.
(i) Information pertaining to the equipment in the process shall include:
(A) Materials of construction;
(B) Piping and instrument diagrams (P&ID's);
(C) Electrical classification;
(D) Relief system design and design basis;
(E) Ventilation system design;
(F) Design codes and standards employed;
(G) Material and energy balances for processes built after May 26, 1992; and,
(H) Safety systems (e.g. interlocks, detection or suppression systems).
(ii) The employer shall document that equipment complies with recognized and generally accepted good engineering practices.
(iii) For existing equipment designed and constructed in accordance with codes, standards, or practices that are no longer in general use, the employer shall determine and document that the equipment is designed, maintained, inspected, tested, and operating in a safe manner.

Appendix 1 - PSM Standard (29 CFR 1910.119)

(e) Process hazard analysis.

(1) The employer shall perform an initial process hazard analysis (hazard evaluation) on processes covered by this standard. The process hazard analysis shall be appropriate to the complexity of the process and shall identify, evaluate, and control the hazards involved in the process. Employers shall determine and document the priority order for conducting process hazard analyses based on a rationale which includes such considerations as extent of the process hazards, number of potentially affected employees, age of the process, and operating history of the process. The process hazard analysis shall be conducted as soon as possible, but not later than the following schedule:

(i) No less than 25 percent of the initial process hazards analyses shall be completed by May 26, 1994;

(ii) No less than 50 percent of the initial process hazards analyses shall be completed by May 26, 1995;

(iii) No less than 75 percent of the initial process hazards analyses shall be completed by May 26, 1996;

(iv) All initial process hazards analyses shall be completed by May 26, 1997.

(v) Process hazards analyses completed after May 26, 1987 which meet the requirements of this paragraph are acceptable as initial process hazards analyses. These process hazard analyses shall be updated and revalidated, based on their completion date, in accordance with paragraph (e)(6) of this section.

(2) The employer shall use one or more of the following methodologies that are appropriate to determine and evaluate the hazards of the process being analyzed.

(i) What-If;
(ii) Checklist;
(iii) What-If/Checklist;
(iv) Hazard and Operability Study (HAZOP):
(v) Failure Mode and Effects Analysis (FMEA);
(vi) Fault Tree Analysis; or
(vii) An appropriate equivalent methodology.

(3) The process hazard analysis shall address:

(i) The hazards of the process;

(ii) The identification of any previous incident which had a likely potential for catastrophic consequences in the workplace;

(iii) Engineering and administrative controls applicable to the hazards and their interrelationships such as appropriate application of detection methodologies to provide early warning of releases. (Acceptable detection methods might include process monitoring and control instrumentation with alarms, and detection hardware such as hydrocarbon sensors.);

(iv) Consequences of failure of engineering and administrative controls;

(v) Facility siting;

(vi) Human factors; and

(vii) A qualitative evaluation of a range of the possible safety and health effects of failure of controls on employees in the workplace.

(4) The process hazard analysis shall be performed by a team with expertise in engineering and process operations, and the team shall include at least one employee who has experience and knowledge specific to the process being evaluated. Also, one member of the team must be knowledgeable in the specific process hazard analysis methodology being used.

(5) The employer shall establish a system to promptly address the team's findings and recommendations; assure that the recommendations are resolved in a timely manner and that the resolution is documented; document what actions are to be taken; complete actions as soon as possible; develop a written schedule of when these actions are to be completed; communicate the actions to operating, maintenance and other employees whose work assignments are in the process and who may be affected by the recommendations or actions.

(6) At least every five (5) years after the completion of the initial process hazard analysis, the process hazard analysis shall be updated and revalidated by a team meeting the requirements in paragraph (e)(4) of this section, to assure that the process hazard analysis is consistent with the current process.

(7) Employers shall retain process hazards analyses and updates or revalidations for each process covered by this section, as well as the documented resolution of recommendations described in paragraph (e)(5) of this section for the life of the process.

(f) Operating procedures

(1) The employer shall develop and implement written operating procedures that provide clear instructions for safely conducting activities involved in each covered process consistent with the process safety information and shall address at least the following elements.

Appendix 1 - PSM Standard (29 CFR 1910.119)

(i) Steps for each operating phase:
(A) Initial startup;
(B) Normal operations;
(C) Temporary operations;
(D) Emergency shutdown including the conditions under which emergency shutdown is required, and the assignment of shutdown responsibility to qualified operators to ensure that emergency shutdown is executed in a safe and timely manner.
(E) Emergency Operations;
(F) Normal shutdown; and,
(G) Startup following a turnaround, or after an emergency shutdown.
(ii) Operating limits:
(A) Consequences of deviation; and
(B) Steps required to correct or avoid deviation.
(iii) Safety and health considerations:
(A) Properties of, and hazards presented by, the chemicals used in the process;
(B) Precautions necessary to prevent exposure, including engineering controls, administrative controls, and personal protective equipment;
(C) Control measures to be taken if physical contact or airborne exposure occurs;
(D) Quality control for raw materials and control of hazardous chemical inventory levels; and,
(E) Any special or unique hazards.
(iv) Safety systems and their functions.
(2) Operating procedures shall be readily accessible to employees who work in or maintain a process.
(3) The operating procedures shall be reviewed as often as necessary to assure that they reflect current operating practice, including changes that result from changes in process chemicals, technology, and equipment, and changes to facilities. The employer shall certify annually that these operating procedures are current and accurate.
(4) The employer shall develop and implement safe work practices to provide for the control of hazards during operations such as lockout/tagout; confined space entry; opening process equipment or piping; and control over entrance into a facility by maintenance, contractor, laboratory, or other support personnel. These safe work practices shall apply to employees and contractor employees.

(g) Training.
   (1) Initial training.
      (i) Each employee presently involved in operating a process, and each employee before being involved in operating a newly assigned process, shall be trained in an overview of the process and in the operating procedures as specified in paragraph (f) of this section. The training shall include emphasis on the specific safety and health hazards, emergency operations including shutdown, and safe work practices applicable to the employee's job tasks.
      (ii) In lieu of initial training for those employees already involved in operating a process on May 26, 1992, an employer may certify in writing that the employee has the required knowledge, skills, and abilities to safely carry out the duties and responsibilities as specified in the operating procedures.
   (2) Refresher training. Refresher training shall be provided at least every three years, and more often if necessary, to each employee involved in operating a process to assure that the employee understands and adheres to the current operating procedures of the process. The employer, in consultation with the employees involved in operating the process, shall determine the appropriate frequency of refresher training.
   (3) Training documentation. The employer shall ascertain that each employee involved in operating a process has received and understood the training required by this paragraph. The employer shall prepare a record which contains the identity of the employee, the date of training, and the means used to verify that the employee understood the training.

(h) Contractors.
   (1) Application. This paragraph applies to contractors performing maintenance or repair, turnaround, major renovation, or specialty work on or adjacent to a covered process. It does not apply to contractors providing incidental services which do not influence process safety, such as janitorial work, food and drink services, laundry, delivery or other supply services.
   (2) Employer responsibilities.
      (i) The employer, when selecting a contractor, shall obtain and evaluate information regarding the contract employer's safety performance and programs.

Appendix 1 - PSM Standard (29 CFR 1910.119)

(ii) The employer shall inform contract employers of the known potential fire, explosion, or toxic release hazards related to the contractor's work and the process.

(iii) The employer shall explain to contract employers the applicable provisions of the emergency action plan required by paragraph (n) of this section.

(iv) The employer shall develop and implement safe work practices consistent with paragraph (f)(4) of this section, to control the entrance, presence and exit of contract employers and contract employees in covered process areas.

(v) The employer shall periodically evaluate the performance of contract employers in fulfilling their obligations as specified in paragraph (h)(3) of this section.

(vi) The employer shall maintain a contract employee injury and illness log related to the contractor's work in process areas.

(3) Contract employer responsibilities.

(i) The contract employer shall assure that each contract employee is trained in the work practices necessary to safely perform his/her job.

(ii) The contract employer shall assure that each contract employee is instructed in the known potential fire, explosion, or toxic release hazards related to his/her job and the process, and the applicable provisions of the emergency action plan.

(iii) The contract employer shall document that each contract employee has received and understood the training required by this paragraph. The contract employer shall prepare a record which contains the identity of the contract employee, the date of training, and the means used to verify that the employee understood the training.

(iv) The contract employer shall assure that each contract employee follows the safety rules of the facility including the safe work practices required by paragraph (f)(4) of this section.

(v) The contract employer shall advise the employer of any unique hazards presented by the contract employer's work, or of any hazards found by the contract employer's work.

(i) Pre-startup safety review.

(1) The employer shall perform a pre-startup safety review for new facilities and for modified facilities when the modification is significant enough to require a change in the process safety information.

(2) The pre-startup safety review shall confirm that prior to the introduction of highly hazardous chemicals to a process:

(i) Construction and equipment is in accordance with design specifications;

(ii) Safety, operating, maintenance, and emergency procedures are in place and are adequate;

(iii) For new facilities, a process hazard analysis has been performed and recommendations have been resolved or implemented before startup; and modified facilities meet the requirements contained in management of change, paragraph (l).

(iv) Training of each employee involved in operating a process has been completed.

(j) Mechanical integrity.

(1) Application. Paragraphs (j)(2) through (j)(6) of this section apply to the following process equipment:

(i) Pressure vessels and storage tanks;

(ii) Piping systems (including piping components such as valves);

(iii) Relief and vent systems and devices;

(iv) Emergency shutdown systems;

(v) Controls (including monitoring devices and sensors, alarms, and interlocks) and,

(vi) Pumps.

(2) Written Procedures. The employer shall establish and implement written procedures to maintain the on-going integrity of process equipment.

(3) Training for process maintenance activities. The employer shall train each employee involved in maintaining the on-going integrity of process equipment in an overview of that process and its hazards and in the procedures applicable to the employee's job tasks to assure that the employee can perform the job tasks in a safe manner.

(4) Inspection and testing.

(i) Inspections and tests shall be performed on process equipment.

(ii) Inspection and testing procedures shall follow recognized and generally accepted good engineering practices.

(iii) The frequency of inspections and tests of process equipment shall be consistent with applicable manufacturers' recommendations and good engineering practices, and more frequently if determined to be necessary by prior operating experience.

(iv) The employer shall document each inspection and test that has been performed on process equipment. The documentation shall identify the date of the inspection or test, the name of the person who performed the inspection or test, the serial number or other identifier of the equipment on which the inspection or test was performed, a description of the inspection or test performed, and the results of the inspection or test.

(5) Equipment deficiencies. The employer shall correct deficiencies in equipment that are outside acceptable limits (defined by the process safety information in paragraph (d) of this section) before further use or in a safe and timely manner when necessary means are taken to assure safe operation.

(6) Quality assurance.

(i) In the construction of new plants and equipment, the employer shall assure that equipment as it is fabricated is suitable for the process application for which they will be used.

(ii) Appropriate checks and inspections shall be performed to assure that equipment is installed properly and consistent with design specifications and the manufacturer's instructions.

(iii) The employer shall assure that maintenance materials, spare parts and equipment are suitable for the process application for which they will be used.

(k) Hot work permit.

(1) The employer shall issue a hot work permit for hot work operations conducted on or near a covered process.

(2) The permit shall document that the fire prevention and protection requirements in 29 CFR 1910.252(a) have been implemented prior to beginning the hot work operations; it shall indicate the date(s) authorized for hot work; and identify the object on which hot work is to be performed. The permit shall be kept on file until completion of the hot work operations.

(l) Management of change.

(1) The employer shall establish and implement written procedures to manage changes (except for "replacements in kind") to process chemicals, technology, equipment, and procedures; and, changes to facilities that affect a covered process.

(2) The procedures shall assure that the following considerations are addressed prior to any change:

(i) The technical basis for the proposed change;
(ii) Impact of change on safety and health;
(iii) Modifications to operating procedures;
(iv) Necessary time period for the change; and,
(v) Authorization requirements for the proposed change.

(3) Employees involved in operating a process and maintenance and contract employees whose job tasks will be affected by a change in the process shall be informed of, and trained in, the change prior to start-up of the process or affected part of the process.

(4) If a change covered by this paragraph results in a change in the process safety information required by paragraph (d) of this section, such information shall be updated accordingly.

(5) If a change covered by this paragraph results in a change in the operating procedures or practices required by paragraph (f) of this section, such procedures or practices shall be updated accordingly.

(m) Incident investigation.

(1) The employer shall investigate each incident which resulted in, or could reasonably have resulted in a catastrophic release of highly hazardous chemical in the workplace.

(2) An incident investigation shall be initiated as promptly as possible, but not later than 48 hours following the incident.

(3) An incident investigation team shall be established and consist of at least one person knowledgeable in the process involved, including a contract employee if the incident involved work of the contractor, and other persons with appropriate knowledge and experience to thoroughly investigate and analyze the incident.

(4) A report shall be prepared at the conclusion of the investigation which includes at a minimum:
(i) Date of incident;
(ii) Date investigation began;
(iii) A description of the incident;
(iv) The factors that contributed to the incident; and,
(v) Any recommendations resulting from the investigation.

(5) The employer shall establish a system to promptly address and resolve the incident report findings and recommendations. Resolutions and corrective actions shall be documented.

Appendix 1 - PSM Standard (29 CFR 1910.119)

(6) The report shall be reviewed with all affected personnel whose job tasks are relevant to the incident findings including contract employees where applicable.

(7) Incident investigation reports shall be retained for five years.

(n) Emergency planning and response. The employer shall establish and implement an emergency action plan for the entire plant in accordance with the provisions of 29 CFR 1910.38(a). In addition, the emergency action plan shall include procedures for handling small releases. Employers covered under this standard may also be subject to the hazardous waste and emergency response provisions contained in 29 CFR 1910.120 (a), (p) and (q).

(o) Compliance audits.

(1) Employers shall certify that they have evaluated compliance with the provisions of this section at least every three years to verify that the procedures and practices developed under the standard are adequate and are being followed.

(2) The compliance audit shall be conducted by at least one person knowledgeable in the process.

(3) A report of the findings of the audit shall be developed.

(4) The employer shall promptly determine and document an appropriate response to each of the findings of the compliance audit, and document that deficiencies have been corrected.

(5) Employers shall retain the two (2) most recent compliance audit reports.

(p) Trade secrets.

(1) Employers shall make all information necessary to comply with the section available to those persons responsible for compiling the process safety information (required by paragraph (d) of this section), those assisting in the development of the process hazard analysis (required by paragraph (e) of this section), those responsible for developing the operating procedures (required by paragraph (f) of this section), and those involved in incident investigations (required by paragraph (m) of this section), emergency planning and response (paragraph (n) of this section) and compliance audits (paragraph (o) of this section) without regard to possible trade secret status of such information.

274 OSHA and EPA PSM Requirements

(2) Nothing in this paragraph shall preclude the employer from requiring the persons to whom the information is made available under paragraph (p)(1) of this section to enter into confidentiality agreements not to disclose the information as set forth in 29 CFR 1910.1200.

(3) Subject to the rules and procedures set forth in 29 CFR 1910.1200(i)(1) through 1910.1200(i)(12), employees and their designated representatives shall have access to trade secret information contained within the process hazard analysis and other documents required to be developed by this standard.

Appendix 2
OSHA List of Highly Hazardous Chemicals, Toxics, and Reactives

Appendix A to § 1910.119—List of Highly Hazardous Chemicals, Toxics and Reactives (Mandatory)

This Appendix contains a listing of toxic and reactive highly hazardous chemicals which present a potential for a catastrophic event at or above the threshold quantity.

CHEMICAL name	CAS*	TQ**
Acetaldehyde	75-07-0	2500
Acrolein (2-Propenal)	107-02-8	150
Acrytyl Chloride	814-68-6	250
Allyl Chloride	107-05-1	1000
Allylamine	107-11-9	1000
Alkylaluminums	Varies	5000
Ammonia, Anhydrous	7664-41-7	10000
Ammonia solutions (>44% ammonia by weight)	7664-41-7	15000
Ammonium Perchlorate	7790-98-9	7500
Ammonium Permanganate	7787-36-2	7500
Arsine (also called Arsenic Hydride)	7784-42-1	100
Bis(Chloromethyl) Ether	542-88-1	100
Boron Trichloride	10294-34-5	2500
Boron Trifluoride	7637-07-2	250
Bromine	7726-95-6	1500
Bromine Chloride	13863-41-7	1500
Bromine Pentafluoride	7789-30-2	2500
Bromine Trifluoride	7787-71-5	15000
3-Bromopropyne (also called Propargyl Bromide)	106-96-7	100
Butyl Hydroperoxide (Tertiary)	75-91-2	5000
Butyl Perbenzoate (Tertiary)	614-45-9	7500
Carbonyl Chloride (see Phosgene)	75-44-5	100
Carbonyl Fluoride	353-50-4	2500
Cellulose Nitrate (concentration >12.6% nitrogen	9004-70-0	2500
Chlorine	7782-50-5	1500
Chlorine Dioxide	10049-04-4	1000
Chlorine Pentrafluoride	13637-63-3	1000
Chlorine Trifluoride	7790-91-2	1000
Chlorodiethylaluminum (also called Diethylaluminum Chloride)	96-10-6	5000
1-Chloro-2, 4-Dinitrobenzene	97-00-7	5000
Chloromethyl Methyl Ether	107-30-2	500
Chloropicrin	76-06-2	500

276 OSHA and EPA PSM Requirements

CHEMICAL name	CAS*	TQ**
Chloropicrin and Methyl Bromide mixture	None	1500
Chloropicrin and Methyl Chloride mixture	None	1500
Cumene Hydroperoxide	80-15-9	5000
Cyanogen	460-19-5	2500
Cyanogen Chloride	506-77-4	500
Cyanuric Fluoride	675-14-9	100
Diacetyl Peroxide (concentraton >70%)	110-22-5	5000
Diazomethane	334-88-3	500
Dibenzoyl Peroxide	94-36-0	7500
Diborane	19287-45-7	100
Dibutyl Peroxide (Tertiary)	110-05-4	5000
Dichloro Acetylene	7572-29-4	250
Dichlorosilane	4109-96-0	2500
Diethylzinc	557-20-0	10000
Diisopropyl Peroxydicarbonate	105-64-6	7500
Dilaluroyl Peroxide	105-74-8	7500
Dimethyldichlorosilane	75-78-5	1000
Dimethylhydrazine, 1,1-	57-14-7	1000
Dimethylamine, Anhydrous	124-40-3	2500
2,4-Dinitroanitine	97-02-9	5000
Ethyl Methyl Ketone Peroxide (also Methyl Ethyl Ketone Peroxide; concentration >60%)	1338-23-4	5000
Ethyl Nitrite	109-95-5	5000
Ethylamine	75-04-7	7500
Ethylene Fluorohydrin	371-62-0	100
Ethylene Oxide	75-21-8	5000
Ethyleneimine	151-56-4	1000
Fluorine	7782-41-4	100
Formaldehyde (Formalin)	50-00-0	1000
Furan	110-00-9	500
Hexafluoroacetone	684-16-2	5000
Hydrochloric Acid, Anhydrous	7647-01-0	5000
Hydrofluoric Acid, Anhydrous	7664-39-3	1000
Hydrogen Bromide	10035-10-6	5000
Hydrogen Chloride	7647-01-0	5000
Hodrogen Cyanide, Anhydrous	74-90-8	1000
Hydrogen Fluoride	7664-39-3	1000
Hydrogen Peroxide (52% by weight or greater	7722-84-1	7500
Hydrogen Selenide	7783-07-5	150
Hydrogen Sulfide	7783=06-4	1500

Appendix 2 - List of Highly Hazardous Chemicals

CHEMICAL name	CAS*	TQ**
Hydroxlamine	7803-49-8	2500
Iron, Pentacarbonyl	13463-40-6	250
Isopropylamine	75-31-0	5000
Ketene	463-51-4	100
Methacrylaldehyde	78-85-3	1000
Methacryloyl Chloride	920-46-7	150
Methacryloyloxyethyl Isocyanate	30674-80-7	100
Methyl Acrylonitrile	126-98-7	250
Methylamine, Anhydrous	74-89-5	1000
Methyl Bromide	74-83-9	2500
Methyl Chloride	74-87-3	15000
Methyl Chloroformate	79-22-1	500
Methyl Ethyl Ketone Peroxide (concentration >60%)	1338-23-4	5000
Methyl Fluoroacetate	453-18-9	100
Methyl Fluorosulfate	421-20-5	100
Methyl Hydrazine	60-34-4	100
Methyl Iodide	74-88-4	7500
Methyl Isocyanate	624-83-9	250
Methyl Mercaptan	74-93-1	5000
Methyl Vinyl Ketone	79-84-4	100
Methyltrichlorosilane	75-79-6	500
Nickel Carbonly (Nickel Tetracarbonyl)	13463-39-3	150
Nitric Acid (94.5% by weight or greater)	7697-37-2	500
Nitric Oxide	10102-43-9	250
Nitroaniline (para Nitroaniline)	100-01-6	5000
Nitromethane	75-52-5	2500
Nitrogen Dioxide	10102-44-0	250
Nitrogen Oxides (NO; NO_2; N204; N203)	10102-44-0	250
Nitrogen Tetroxide (also called Nitrogen Peroxide)	10544-72-6	250
Nitrogen Trifluoride	7783-54-2	5000
Nitrogen Trioxide	10544-73-7	250
Oleum (65% to 80% by weight; also called Fuming Sulfuric Acid	8014-94-7	1000
Osmium Tetroxide	20816-12-0	100
Oxygen Difluoride (Fluorine Monoxide)	7783-41-7	100
Ozone	10028-15-6	100
Pentaborane	19624-22-7	100

278 OSHA and EPA PSM Requirements

CHEMICAL name	CAS*	TQ**
Peracetic Acid (concentration >60% Acetic Acid; also called Peroxyacetic Acid)	79-21-0	1000
Perchloric Acid (concentration >60% by weight)	7601-90-3	5000
Perchloromethyl Mercaptan	594-42-3	150
Perchloryl Fluoride	7616-94-6	5000
Peroxyacetic Acid (concentration >60% by Acetic Acid; also called Paracetic Acid)	79-21-0	1000
Phosgene (also called Carbonyl Chloride)	75-44-5	100
Phosphine (Hydrogen Phosphide)	7803-51-2	100
Phosphorus Oxychloride (also called Phosphoryl Chloride)	10025-87-3	1000
Phosphorus Trichloride	7719-12-2	1000
Phosphoryl Chloride (also called Phosphorus Oxychloride)	10025-87-3	1000
Propargyl Bromide	106-96-7	100
Propyl Nitrate	627-3-4	100
Sarin	107-44-8	100
Selenium Hexafluoride	7783-79-1	1000
Stibine (Antimony Hydride)	7803-52-3	500
Sulfur Dioxide (liquid)	7446-09-5	1000
Sulfur Pentafluoride	5714-22-7	250
Sulfur Tetrafluoride	7783-60-0	250
Sulfur Trioxide (also called Sulfuric Anhydride)	7446-11-9	1000
Sulfuric Anhydride (also called Sulfur Trioxide)	7446-11-9	1000
Tellurium Hexafluoride	7783-80-4	250
Tetrafluoroethylene	116-14-3	5000
Tetrafluorohydrazine	10036-47-2	5000
Tetramethyl Lead	75-74-1	1000
Thionyl Chloride	7719-09-7	250
Trichloro (chloromethyl) Silane	1558-25-4	100
Trichloro (dichlorophenyl) Silane	27137-85-5	2500
Trichlorosilane	10025-78-2	5000
Triflurochloroethylene	79-38-9	10000
Trimethyoxysilane	2487-90-3	1500

*Chemical Abstract Service Number.
**Threshold Quantity in Pounds (Amount necessary to be covered by this standard).

Appendix 3

Block Flow Diagram and Simplified Process Flow Diagram

Appendix B to § 1910.119—Block Flow Diagram and Simplified Process Flow Diagram (Nonmandatory)

Example of a Block Flow Diagram

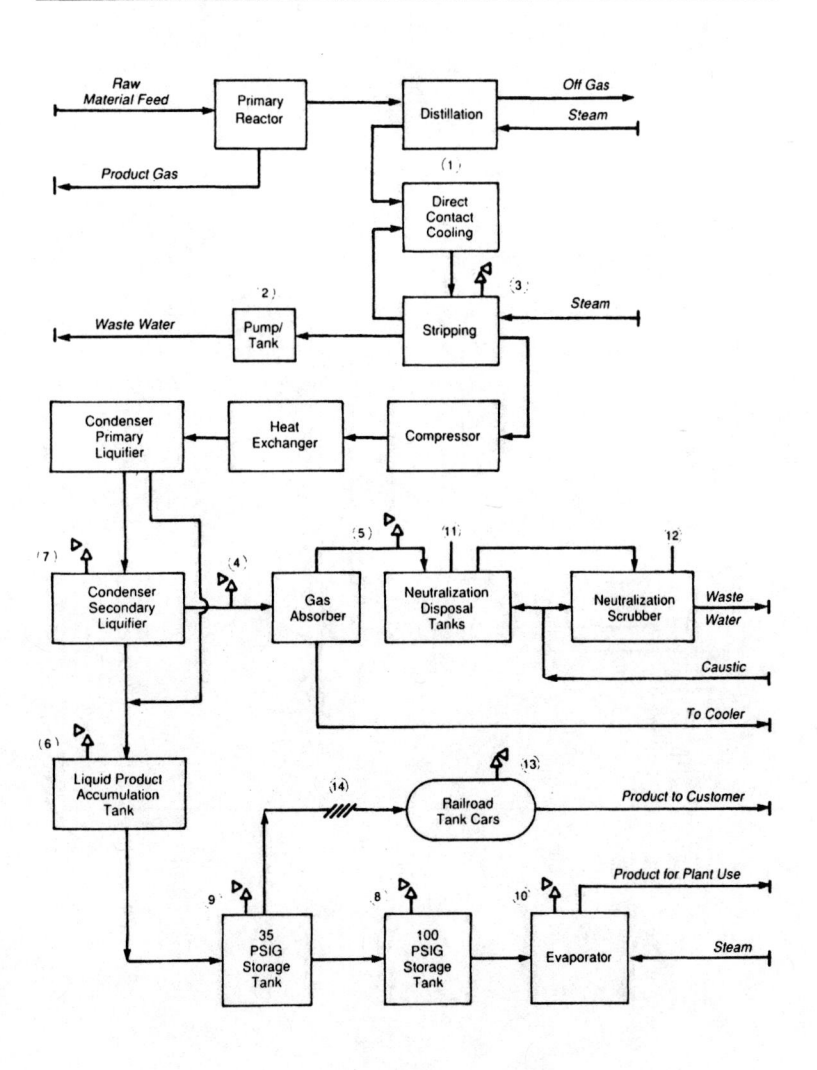

280 OSHA and EPA PSM Requirements

Example of a Simplified Process Flow Diagram

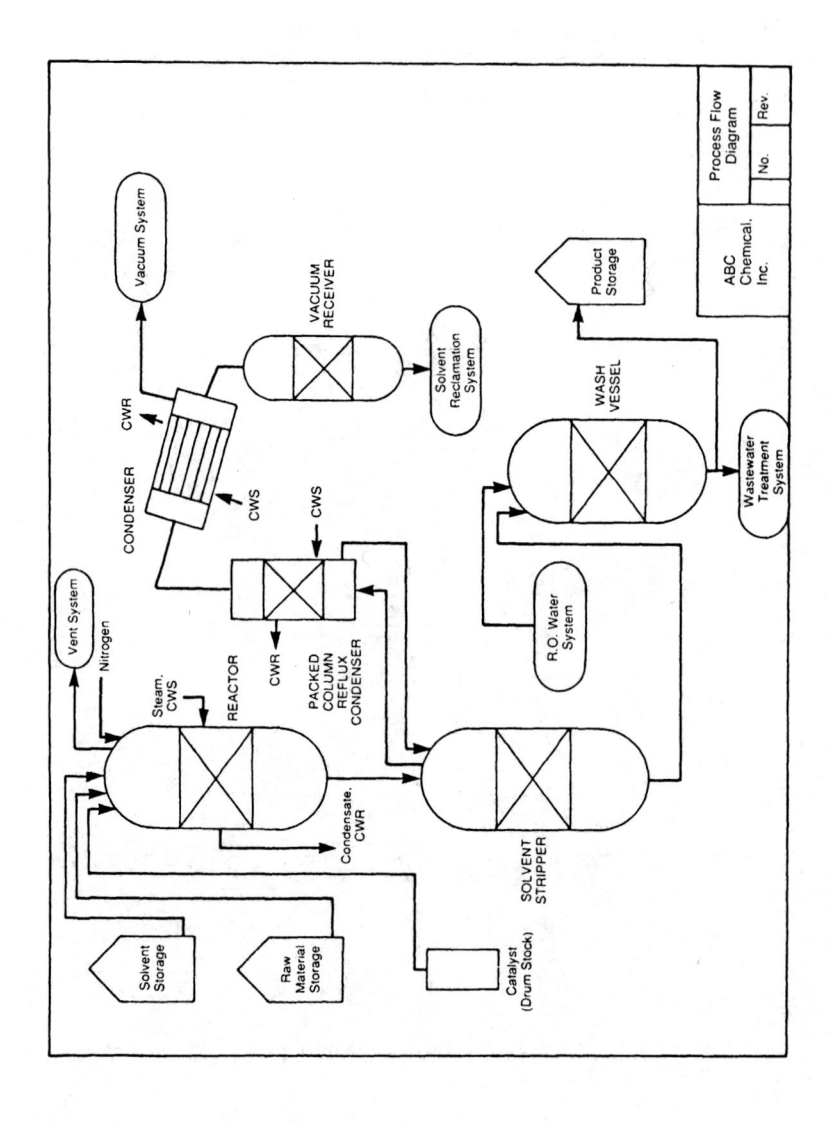

Appendix 4

Sources of Further Information on Process Safety Management

1. Center for Chemical Process Safety, American Institute of Chemical Engineers, 345 East 47th Street, New York, NY 10017, (212) 705-7319.

2. "Guidelines for Hazard Evaluation Procedures," American Institute of Chemical Engineers; 345 East 47th Street, New York, NY 10017.

3. "Guidelines for Technical Management of Chemical Process Safety," Center for Chemical Process Safety of the American Institute of Chemical Engineers; 345 East 47th Street, New York, NY 10017.

4. "Evaluating Process Safety in the Chemical Industry," Chemical Manufacturers Association; 2501 M Street NW, Washington, DC 20037.

5. "Safe Warehousing of Chemicals," Chemical Manufacturers Association; 2501 M Street NW, Washington, DC 20037.

6. "Management of Process Hazards," American Petroleum Institute (API Recommended Practice 750); 1220 L Street, N.W., Washington, D.C. 20005.

7. "Improving Owner and Contractor Safety Performance," American Petroleum Institute (API Recommended Practice 2220); API, 1220 L Street N.W., Washington, D.C. 20005.

8. Chemical Manufacturers Association (CMA's Manager Guide), First Edition, September 1991; CMA, 2501 M Street, N.W., Washington, D.C. 20037.

9. "Improving Construction Safety Performance," Report A-3, The Business Roundtable; The Business Roundtable, 200 Park Avenue, New York, NY 10166.
(Report includes criteria to evaluate contractor safety performance and criteria to enhance contractor safety performance).

10. "Recommended Guidelines for Contractor Safety and Health," Texas Chemical Council; Texas Chemical Council, 1402 Nueces Street, Austin, TX 78701-1534.

11. "Loss Prevention in the Process Industries," Volumes I and II; Frank P. Lees, Butterworth; London 1983.

12. "Safety and Health Program Management Guidelines," 1989; U.S. Department of Labor, Occupational Safety and Health Administration.

13. "Safety and Health Guide for the Chemical Industry," 1986, (OSHA 3091); U.S. Department of Labor, Occupational Safety and Health Administration; 200 Constitution Avenue, N.W., Washington, D.C. 20210.

14. "Review of Emergency Systems," June 1988; U.S. Environmental Protection Agency (EPA), Office of Solid Waste and Emergency Response, Washington, DC 20460.

15. "Technical Guidance for Hazards Analysis, Emergency Planning for Extremely Hazardous Substances," December 1987; U.S. Environmental Protection Agency (EPA), Federal Emergency Management Administration (FEMA) and U.S. Department of Transportation (DOT), Washington, DC 20460.

16. "Accident Investigation ... A New Approach," 1983, National Safety Council; 444 North Michigan Avenue, Chicago, IL 60611-3991.

Appendix 4 - PSM Information Sources

17. "Fire & Explosion Index Hazard Classification Guide," 6th Edition, May 1987, Dow Chemical Company; Midland, Michigan 48674.

18. "Chemical Exposure Index," May 1988, Dow Chemical Company; Midland, Michigan 48674.

[Source: 57 Fed. Reg. 6403 (Feb. 24, 1992).]

Appendix 5
Directory of Consultation Programs
OSHA Consultation Directory

Consultation programs provide free services to employers who request help in identifying and correcting specific hazards, want to improve their safety and health programs, and/or need further assistance in training and education. Funded by OSHA and delivered by well-trained professional staff of state governments, consultation services are comprehensive, and include an appraisal of all workplace hazards, practices, and job safety and health programs; conferences and agreements with management; assistance in implementing recommendations; and a follow-up appraisal to ensure that any required corrections are made. For more information on consultation programs, contact the appropriate office in your state listed below.

State	Telephone
Alabama	(205) 348-3033
Alaska	(907) 264-2599
Arizona	(602) 255-5795
Arkansas	(501) 682-4522
California	(415) 703-4441
Colorado	(303) 491-6151
Connecticut	(203) 566-4550
Delaware	(302) 577-3908
District of Columbia	(202) 576-6339
Florida	(904) 488-3044
Georgia	(404) 894-8274
Guam	(671) 646-9244
Hawaii	(808) 548-4155
Idaho	(208) 385-3283
Illinois	(312) 814-2337
Indiana	(317) 232-2688
Iowa	(515) 281-5352
Kansas	(913) 296-4386
Kentucky	(502) 564-6895
Louisiana	(504) 342-9601
Maine	(207) 289-6460
Maryland	(301) 333-4218
Massachusetts	(617) 727-3463
Michigan	(517) 335-8250 *(H)*
	(517) 322-1809 *(S)*
Minnesota	(612) 297-2393
Mississippi	(601) 987-3981
Missouri	(314) 751-3403
Montana	(406) 444-6401

H - Health   S - Safety

Appendix 5 - Consultation Programs

Nebraska	(402) 471-4717
Nevada	(702) 486-5016
New Hampshire	(603) 271-2024
New Jersey	(609) 292-7036
New Mexico	(505) 827-2885
New York	(518) 457-2481
North Carolina	(919) 733-3949
North Dakota	(701) 221-5188
Ohio	(614) 644-2631
Oklahoma	(405) 528-1500
Oregon	(503) 378-3272
Pennsylvania	(304) 558-7890
Puerto Rico	(809) 754-2171
Rhode Island	(401) 277-2438
South Carolina	(803) 734-9599
South Dakota	(605) 688-4101
Tennessee	(615) 741-7036
Texas	(512) 440-3834
Utah	(801) 530-6868
Vermont	(802) 828-2765
Virginia	(804) 786-6613
Virgin Islands	(809) 772-1315
Washington	(206) 586-0963
West Virginia	(304) 348-7890
Wisconsin	(608) 266-5879 *(H)*
	(414) 521-5063 *(S)*
Wyoming	(307) 777-7786

H - Health   S - Safety

Appendix 6
States With Approved OSHA Programs

States with Approved Plans

States administering their own occupational safety and health programs through plans approved under section 18(b) of the Occupational Safety and Health Act of 1970 must adopt standards and enforce requirements that are at least as effective as federal requirements.

There are currently 25 state plan states; 23 cover the private and public (state and local government) sections and 2 cover the public sector only (Connecticut and New York).

COMMISSIONER
Alaska Department of Labor
1111 West 8th Street
Room 306
Juneau, AK 99801
(907) 465-2700

DIRECTOR
Industrial Commission
 of Arizona
800 W. Washington
Phoenix, AZ 85007
(602) 542-5795

DIRECTOR
California Department
 of Industrial Relations
455 Golden Gate Avenue
4th Floor
San Francisco, CA 94102
(415) 703-4590

COMMISSIONER
Connecticut Department
 of Labor
200 Folly Brook Boulevard
Wethersfield, CT 06109
(203) 566-5123

DIRECTOR
Hawaii Department of Labor
 and Industrial Relations
830 Punchbowl Street
Honolulu, HI 96813
(808) 586-8844

COMMISSIONER
Indiana Department of Labor
State Office Building
402 West Washington Street
Room W195
Indianapolis, IN 46204
(317) 232-2378

COMMISSIONER
Iowa Division of Labor Services
1000 E. Grand Avenue
Des Moines, IA 50319
(515) 281-3447

ACTING COMMISSIONER
for Workplace Standards
Kentucky Labor Cabinet
1049 U.S. Highway, 127 South
Frankfort, KY 40601
(502) 564-3070

COMMISSIONER
Maryland Division of Labor
 and Industry
Department of Licensing
 and Regulation
501 St. Paul Place, 2nd Floor
Baltimore, MD 21202-2272
(301) 333-4179

DIRECTOR
Michigan Department of Labor
Victor Office Center
201 N. Washington Square
P.O. Box 30015
Lansing, MI 48933
(517) 373-9600

Appendix 6 - State Programs

DIRECTOR
Michigan Department of Public Health
3423 North Logan Street
Box 30195
Lansing, MI 48909
(517) 335-8022

COMMISSIONER
Minnesota Department of Labor and Industry
443 Lafayette Road
St. Paul, MN 55155
(612) 296-2342

DIRECTOR
Nevada Department of Industrial Relations
Division of Occupational Safety and Health
Capitol Complex
1370 S. Curry Street
Carson City, NV 89710
(702) 687-3032

SECRETARY
New Mexico Environment Dept.
Occupational Health and Safety Bureau
1190 St. Francis Drive
P.O. Box 26110
Santa Fe, NM 87502
(505) 827-2850

COMMISSIONER
New York Department of Labor
State Office Building
Campus 12-Room 457
Albany, NY 12240
(518) 457-2741

COMMISSIONER
North Carolina Department of Labor
4 West Edenton Street
Raleigh, NC 27601
(919) 733-0360

ADMINISTRATOR
Oregon Occupational Safety and Health Division
Oregon Department of Insurance and Finance, Room 160
21 Labor and Industries Building
Summer and Chemeketa Streets, NE
Salem, OR 97310
(503) 378-3272

SECRETARY
Puerto Rico Department of Labor and Human Resources
Prudencio Rivera Martinez Building
505 Munoz Rivera Avenue
Hato Rey, PR 00918
(809) 754-2119

COMMISSIONER
South Carolina Department of Labor
3600 Forest Drive
P.O. Box 11329
Columbia, SC 29211-1329
(803) 734-9594

COMMISSIONER
Tennessee Department of Labor
501 Union Building
Suite "A" - 2nd Floor
Nashville, TN 37243-0655
(615) 741-2582

ADMINISTRATOR
Utah Occupational Safety and Health
160 East 300 South
P.O. Box 5800
Salt Lake City, UT 84110-5800
(801) 530-6900

COMMISSIONER
Vermont Department of Labor and Industry
120 State Street
Montpelier, VT 05620
(802) 828-2288

288 OSHA and EPA PSM Requirements

COMMISSIONER
Virgin Islands Department
 of Labor
2131 Hospital Street
Box 890
Christiansted
St. Croix, VI 00840-4666
(809) 773-1994

COMMISSIONER
Virginia Department of Labor
 and Industry
Powers-Taylor Building
13 S. 13th Street
Richmond, VA 23219
(804) 786-2376

DIRECTOR
Washington Department
 of Labor and Industries
General Administration Building
Room 334-AX-31
Olympia, WA 98504-0631
(206) 753-6307

DIRECTOR
Department of Employment
Division of Employment Affairs
Occupational Safety and Health
 Administration
Herschler Building, 2nd Floor
 East
122 West 25th Street
Cheyenne, WY 82002
(307) 777-7672

Appendix 7

EPA List of Regulated Substances and Thresholds for Accidental Release Prevention

PART 68 - CHEMICAL ACCIDENT PREVENTION PROVISIONS

Subpart A - General
  68.1 Scope.
  68.3 Definitions.

Subpart B - Risk Management Plan Requirements (Reserved)

Subpart C - Regulated Substances for Accidental Release Prevention
  68.100 Purpose.
  68.115 Threshold determination.
  68.120 Petition process.
  68.125 Exemptions.
  68.130 List of substances.

[Authority: 42 U.S.C. 7412(r), 7601.]

Subpart A - General

68.1 Scope.

This Part sets forth the list of regulated substances and thresholds, the petition process for adding or deleting substances to the list of regulated substances, the requirements for owners or operators of stationary sources concerning the prevention of accidental releases,

290 OSHA and EPA PSM Requirements

and the State accidental release prevention programs approved under section 112(r). The list of substances, threshold quantities, and accident prevention regulations promulgated under this part do not limit in any way the general duty provisions under section 112(r)(1).

68.3 Definitions.

For the purposes of this Part:

Accidental release means an unanticipated emission of a regulated substance or other extremely hazardous substance into the ambient air from a stationary source.

Administrator means the administrator of the U.S. Environmental Protection Agency.

Article means a manufactured item, as defined under 29 CFR 1910.1200(b), that is formed to a specific shape or design during manufacture, that has end use functions dependent in whole or in part upon the shape or design during end use, and that does not release or otherwise result in exposure to a regulated substance under normal conditions of processing and use.

CAS means the Chemical Abstracts Service.

DOT means the United States Department of Transportation.

Process means any activity involving a regulated substance including any use, storage, manufacturing, handling, or on-site movement of such substances, or combination of these activities. For the purposes of this definition, any group of vessels that are interconnected, or separate vessels that are located such that a regulated substance could be involved in a potential release, shall be considered a single process.

Regulated substance is any substance listed pursuant to section 112(r)(3) of the Clean Air Act as amended, in 68.130.

Stationary source means any buildings, structures, equipment, installations, or substance emitting stationary activities which belong to the same industrial group, which are located on one or more contiguous properties, which are under the control of the same person (or persons under common control), and from which an accidental release may occur. A stationary source includes transportation containers that are no longer under active shipping papers and transportation containers that are connected to equipment at the stationary source for the purposes of temporary storage, loading, or unloading. The term stationary source does not

Appendix 7 - EPA Regulated Substances

apply to transportation, including the storage incident to transportation, of any regulated substance or any other extremely hazardous substance under the provisions of this part, provided that such transportation is regulated under 49 CFR parts 192, 193, or 195. Properties shall not be considered contiguous solely because of a railroad or gas pipeline right-of-way.

Threshold quantity means the quantity specified for regulated substances pursuant to section 112(r)(5) of the Clean Air Act as amended, listed in 68.130 and determined to be present at a stationary source as specified in 68.115 of this Part.

Vessel means any reactor, tank, drum, barrel, cylinder, vat, kettle, boiler, pipe, hose, or other container.

Subpart B - Risk Management Plan Requirements (Reserved)

Subpart C - Regulated Substances for Accidental Release Prevention

68.100 Purpose.

This subpart designates substances to be listed under section 112(r)(3), (4), and (5) of the Clean Air Act, as amended, identifies their threshold quantities, and establishes the requirements for petitioning to add or delete substances from the list.

68.115 Threshold determination.

(a) A threshold quantity of a regulated substance listed in 68.130 is present at a stationary source if the total quantity of the regulated substance contained in a process exceeds the threshold.

(b) For the purposes of determining whether more than a threshold quantity of a regulated substance is present at the stationary source, the following exemptions apply:

(1) Concentrations of a regulated toxic substance in a mixture. If a regulated substance is present in a mixture and the concentration of the substance is below one percent by weight of the mixture, the amount of the substance in the mixture need not be considered when determining whether more than a threshold quantity is present at the stationary source. Except for oleum, toluene 2,4-diisocyanate, toluene 2,6-diisocyanate, and toluene diisocyanate (unspecified isomer), if the concentration of the regulated substance in the mixture is one percent or greater by

weight, but the owner or operator can demonstrate that the partial pressure of the regulated substance in the mixture (solution) under handling or storage conditions in any portion of the process is less than 10 millimeters of mercury (mm Hg), the amount of the substance in the mixture in that portion of the process need not be considered when determining whether more than a threshold quantity is present at the stationary source. The owner or operator shall document this partial pressure measurement or estimate.

(2) Concentrations of a regulated flammable substance in a mixture. If a regulated substance is present in a mixture and the concentration of the substance is below one percent by weight of the mixture, the mixture need not be considered when determining whether more than a threshold quantity of the regulated substance is present at the stationary source. If the concentration of the regulated substance in the mixture is one percent or greater by weight, then, for purposes of determining whether more than a threshold quantity is present at the stationary source, the entire weight of the mixture shall be treated as the regulated substance unless the owner or operator can demonstrate that the mixture itself does not meet the criteria for flammability of flash point below 73°F (22.8°C) and boiling point below 100°F (37.8°C). The owner or operator shall document these flash point and boiling point measurements or estimates.

(3) Concentrations of a regulated explosive substance in a mixture. Mixtures of Division 1.1 explosives listed in 49 CFR 172.101 (Hazardous Materials Table) and other explosives need not be included when determining whether a threshold quantity is present in a process, when the mixture is intended to be used on-site in a non-accidental release in a manner consistent with applicable BATF regulations. Other mixtures of Division 1.1 explosives listed in 49 CFR 172.101 and other explosives shall be included in determining whether more than a threshold quantity is present in a process if such mixtures would be treated as Division 1.1 explosives under 49 CFR Parts 172 and 173.

(4) Articles. Regulated substances contained in articles need not be considered when determining whether more than a threshold quantity is present at the stationary source.

(5) Uses. Regulated substances, when in use for the following purposes, need not be included in determining whether more than a threshold quantity is present at the stationary source:

    (i) Use as a structural component of the stationary source;
    (ii) Use of products for routine janitorial maintenance;

(iii) Use by employees of foods, drugs, cosmetics, or other personal items containing the regulated substance; and

(iv) Use of regulated substances present in process water or non-contact cooling water as drawn from the environment or municipal sources, or use of regulated substances present in air used either as compressed air or as part of combustion.

(6) Activities in Laboratories. If a regulated substance is manufactured, processed, or used in a laboratory at a stationary source under the supervision of a technically qualified individual as defined in s720.3(ee) of this chapter, the quantity of the substance need not be considered in determining whether a threshold quantity is present. This exemption does not apply to:

(i) Specialty chemical production;

(ii) Manufacture, processing, or use of substances in pilot plant scale operations; and

(iii) Activities conducted outside the laboratory.

68.120 Petition process.

(a) Any person may petition the Administrator to modify, by addition or deletion, the list of regulated substances identified in 68.130. Based on the information presented by the petitioner, the Administrator may grant or deny a petition.

(b) A substance may be added to the list if, in the case of an accidental release, it is known to cause or may be reasonably anticipated to cause death, injury, or serious adverse effects to human health or the environment.

(c) A substance may be deleted from the list if adequate data on the health and environmental effects of the substance are available to determine that the substance, in the case of an accidental release, is not known to cause and may not be reasonably anticipated to cause death, injury, or serious adverse effects to human health or the environment.

(d) No substance for which a national primary ambient air quality standard has been established shall be added to the list. No substance regulated under Title VI of the Clean Air Act, as amended, shall be added to the list.

(e) The burden of proof is on the petitioner to demonstrate that the criteria for addition and deletion are met. A petition will be denied if this demonstration is not made.

(f) The Administrator will not accept additional petitions on the same substance following publication of a final notice of the decision to grant or deny a petition, unless new data becomes available that could significantly affect the basis for the decision.

(g) Petitions to modify the list of regulated substances must contain the following:

(1) Name and address of the petitioner and a brief description of the organization(s) that the petitioner represents, if applicable;

(2) Name, address, and telephone number of a contact person for the petition;

(3) Common chemical name(s), common synonym(s), Chemical Abstracts Service number, and chemical formula and structure;

(4) Action requested (add or delete a substance);

(5) Rationale supporting the petitioner's position; that is, how the substance meets the criteria for addition and deletion. A short summary of the rationale must be submitted along with a more detailed narrative; and (6) Supporting data; that is, the petition must include sufficient information to scientifically support the request to modify the list. Such information shall include:

(i) A list of all support documents;

(ii) Documentation of literature searches conducted, including, but not limited to, identification of the database(s) searched, the search strategy, dates covered, and printed results;

(iii) Effects data (animal, human, and environmental test data) indicating the potential for death, injury, or serious adverse human and environmental impacts from acute exposure following an accidental release; printed copies of the data sources, in English, should be provided; and

(iv) Exposure data or previous accident history data, indicating the potential for serious adverse human health or environmental effects from an accidental release. These data may include, but are not limited to, physical and chemical properties of the substance, such as vapor pressure; modeling results, including data and assumptions used and model documentation; and historical accident data, citing data sources.

(h) Within 18 months of receipt of a petition, the Administrator shall publish in the Federal Register a notice either denying the petition or granting the petition and proposing a listing.

68.125 Exemptions.

Agricultural nutrients. Ammonia used as an agricultural nutrient, when held by farmers, is exempt from all provisions of this part.

68.130 List of substances.

(a) Explosives listed by DOT as Division 1.1 in 49 CFR 172.101 are covered under section 112(r) of the Clean Air Act. The threshold quantity for explosives is 5,000 pounds.

(b) Regulated toxic and flammable substances under section 112(r) of the Clean Air Act are the substances listed in Tables 1, 2, 3, and 4. Threshold quantities for listed toxic and flammable substances are specified in the tables.

(c) The basis for placing toxic and flammable substances on the list of regulated substances are explained in the notes to the list.

296 OSHA and EPA PSM Requirements

Table 1 to 68.130. - List of Regulated Toxic Substances and Threshold
Quantities for Accidental Release Prevention
(Alphabetical Order--77 Substances)

Chemical name	CAS No.	Threshold quantity (lbs)	Basis for listing
Acrolein (2-Propenal)	107-02-8	5,000	b
Acrylonitrile (2-Propenenitrile)	107-13-1	20,000	b
Acrylyl chloride (2-Propenoyl chloride)	814-68-6	5,000	b
Allyl alcohol (2-Propen-1-ol)	107-18-61	15,000	b
Allylamine (2-Propen-1-amine)	107-11-9	10,000	b
Ammonia (anhydrous)	7664-41-7	10,000	a, b
Ammonia (conc 20% or greater)	7664-41-7	20,000	a, b
Arsenous trichloride	7784-34-1	15,000	b
Arsine	7784-42-1	1,000	b
Boron trichloride (Borane, trichloro-)	10294-34-5	5,000	b
Boron trifluoride (Borane, trifluoro-)	7637-07-2	5,000	b
Boron trifluoride compound with methyl ether (1:1) (Boron, trifluoro(oxybis(metane))-, T-4-	353-42-4	15,000	b
Bromine	7726-95-6	10,000	a, b
Carbon disulfide	75-15-0	20,000	b
Chlorine	7782-50-5	2,500	a, b
Chlorine dioxide (Chlorine oxide (ClO_2))	10049-04-4	1,000	c
Chloroform (Methane, trichloro-)	67-66-3	20,000	b
Chloromethyl ether (Methane, oxybis(chloro-)	542-88-1	1,000	b
Chloromethyl methyl ether (Methane, chloromethoxy-)	107-30-2	5,000	b
Crotonaldehyde (2-Butenal)	4170-30-3	20,000	b
Crotonaldehyde, (E)- (2-Butenal, (E)-)	123-73-9	20,000	b
Cyanogen chloride	506-77-4	10,000	c
Cyclohexylamine (Cyclohexanamine)	108-91-8	15,000	b
Diborane	19287-45-7	2,500	b
Dimethyldichlorosilane (Silane, dichlorodimethyl-)	75-78-5	5,000	b
1,1-Dimethylhydrazine (Hydrazine, 1,1-dimethyl-)	57-14-7	15,000	b
Epichlorohydrin (Oxirane, (chloromethyl)-)	106-89-8	20,000	b
Ethylenediamine (1,2-Ethanediamine)	107-15-3	20,000	b
Ethyleneimine (Aziridine)	151-56-4	10,000	b
Ethylene oxide (Oxirane)	75-21-8	10,000	a, b
Fluorine	7782-41-4	1,000	b
Formaldehyde (solution)	50-00-0	15,000	b
Furan	110-00-9	5,000	b
Hydrazine	302-01-2	15,000	b
Hydrochloric acid (conc 30% or greater)	7647-01-0	15,000	d
Hydrocyanic acid	74-90-8	2,500	a, b
Hydrogen chloride (anhydrous)			

Appendix 7 - EPA Regulated Substances

Chemical name	CAS No.	Threshold quantity (lbs)	Basis for listing
(Hydrochloric acid)	7647-01-0	5,000	a
Hydrogen fluoride/Hydrofluoric acid (conc 50% or greater) (Hydrofluoric acid)	7664-39-3	1,000	a, b
Hydrogen selenide	7783-07-5	500	b
Hydrogen sulfide	7783-06-4	10,000	a, b
Iron, pentacarbonyl- (Iron carbonyl (Fe(CO)$_5$), (TB-5-11)-)	13463-40-6	2,500	b
Isobutyronitrile (Propanenitrile, 2-methyl-)	78-82-0	20,000	b
Isopropyl chloroformate (Carbonochloridic acid, 1-methylethyl ester)	108-23-6	15,000	b
Methacrylonitrile (2-Propenenitrile, 2-methyl-)	126-98-7	10,000	b
Methyl chloride (Methane, chloro-)	74-87-3	10,000	a
Methyl chloroformate (Carbonochloridic acid, methylester)	79-22-1	5,000	b
Methyl hydrazine (Hydrazine, methyl-)	60-34-4	15,000	b
Methyl isocyanate (Methane, isocyanato-)	624-83-9	10,000	a, b
Methyl mercaptan (Methanethiol)	74-93-1	10,000	b
Methyl thiocyanate (Thiocyanic acid, methyl ester)	556-64-9	20,000	b
Methyltrichlorosilane (Silane, trichloromethyl-)	75-79-6	5,000	b
Nickel carbonyl	13463-39-3	1,000	b
Nitric acid (conc 80% or greater)	7697-37-2	15,000	b
Nitric oxide (Nitrogen oxide (NO))	10102-43-9	10,000	b
Oleum (Fuming Sulfuric acid) (Sulfuric acid, mixture with sulfur trioxide)[1]	8014-95-7	10,000	e
Peracetic acid (Ethaneperoxoic acid)	79-21-0	10,000	b
Perchloromethylmercaptan (Methanesulfenyl chloride, trichloro-)	594-42-3	10,000	b
Phosgene (Carbonic dichloride)	75-44-5	500	a, b
Phosphine	7803-51-2	5,000	b
Phosphorus oxychloride (Phosphoryl chloride)	10025-87-3	5,000	b
Phosphorus trichloride (Phosphorous trichloride)	7719-12-2	15,000	b
Piperidine	110-89-4	15,000	b
Propionitrile (Propanenitrile)	107-12-0	10,000	b
Propyl chloroformate (Carbonochloridic acid, propylester)	109-61-5	15,000	b
Propyleneimine (Aziridine, 2-methyl-)	75-55-8	10,000	b
Propylene oxide (Oxirane, methyl-)	75-56-9	10,000	b
Sulfur dioxide (anhydrous)	7446-09-5	5,000	a, b

298 OSHA and EPA PSM Requirements

Chemical name	CAS No.	Threshold quantity (lbs)	Basis for listing
Sulfur tetrafluoride (Sulfur fluoride (SF4), (T-4)-)	7783-60-0	2,500	b
Sulfur trioxide	7446-11-9	10,000	a, b
Tetramethyllead (Plumbane, tetramethyl-)	75-74-1	10,000	b
Tetranitromethane (Methane, tetranitro-)	509-14-8	10,000	b
Titanium tetrachloride (Titanium chloride (TiCl4) (T-4)-)	7550-45-0	2,500	b
Toluene 2,4-diisocyanate (Benzene, 2,4-diisocyanato-1-methyl-)[1]	584-84-9	10,000	a
Toluene 2,6-diisocyanate (Benzene, 1,3-diisocyanato-2-methyl-)[1]	91-08-7	10,000	a
Toluene diisocyanate (unspecified isomer) (Benzene, 1,3-diisocyanatomethyl-)[1]	26471-62-5	10,000	a
Trimethylchlorosilane (Silane, chlorotrimethyl-)	75-77-4	10,000	b
Vinyl acetate monomer (Acetic acid ethenyl ester)	108-05-4	15,000	b

[1] The mixture exemption in 68.115(b)(1) does not apply to the substance.

Note: Basis for Listing:
a Mandated for listing by Congress.
b On EHS list, vapor pressure 10 mmHg or greater.
c Toxic gas.
d Toxicity of hydrogen chloride, potential to release hydrogen chloride, and history of accidents.
e Toxicity of sulfur trioxide and sulfuric acid, potential to release sulfur trioxide, and history of accidents.

Appendix 7 - EPA Regulated Substances

Table 2 to 68.130. - List of Regulated Toxic Substances and Threshold Quantities for Accidental Release Prevention (CAS Number Order--77 Substances)

CAS No.	Chemical name	Threshold quantity (lbs)	Basis for listing
50-00-0	Formaldehyde (solution)	15,000	b
57-14-7	1,1-Dimethylhydrazine (Hydrazine, 1,1-dimethyl-)	15,000	b
60-34-4	Methyl hydrazine (Hydrazine, methyl-)	15,000	b
67-66-3	Chloroform (Methane, trichloro-)	20,000	b
74-87-3	Methyl chloride (Methane, chloro-)	10,000	a
74-90-8	Hydrocyanic acid	2,500	a, b
74-93-1	Methyl mercaptan (Methanethiol)	10,000	b
75-15-0	Carbon disulfide	20,000	b
75-21-8	Ethylene oxide (Oxirane)	10,000	a, b
75-44-5	Phosgene (Carbonic dichloride)	500	a, b
75-55-8	Propyleneimine (Aziridine, 2-methyl-)	10,000	b
75-56-9	Propylene oxide (Oxirane, methyl-)	10,000	b
75-74-1	Tetramethyllead (Plumbane, tetramethyl-)	10,000	b
75-77-4	Trimethylchlorosilane (Silane, chlorotrimethyl-)	10,000	b
75-78-5	Dimethyldichlorosilane (Silane, dichlorodimethyl-)	5,000	b
75-79-6	Methyltrichlorosilane (Silane, trichloromethyl-)	5,000	b
78-82-0	Isobutyronitrile (Propanenitrile, 2-methyl-)	20,000	b
79-21-0	Peracetic acid (Ethaneperoxoic acid)	10,000	b
79-22-1	Methyl chloroformate (Carbonochloridic acid, methylester)	5,000	b
91-08-7	Toluene 2,6-diisocyanate (Benzene, 1,3-diisocyanato-2-methyl-)[1]	10,000	a
106-89-8	Epichlorohydrin (Oxirane, (chloromethyl)-)	20,000	b
107-02-8	Acrolein (2-Propenal)	5,000	b
107-11-9	Allylamine (2-Propen-1-amine)	10,000	b
107-12-0	Propionitrile (Propanenitrile)	10,000	b
107-13-1	Acrylonitrile (2-Propenenitrile)	20,000	b
107-15-3	Ethylenediamine (1,2-Ethanediamine)	20,000	b
107-18-6	Allyl alcohol (2-Propen-1-ol)	15,000	b
107-30-2	Chloromethyl methyl ether (Methane, chloromethoxy-)	5,000	b
108-05-4	Vinyl acetate monomer (Acetic acid ethenyl ester)	15,000	b
108-23-6	Isopropyl chloroformate (Carbonochloridic acid, 1-methylethyl ester)	15,000	b
108-91-8	Cyclohexylamine (Cyclohexanamine)	15,000	b

300 OSHA and EPA PSM Requirements

CAS No.	Chemical name	Threshold quantity (lbs)	Basis for listing
109-61-5	Propyl chloroformate (Carbonochloridic acid, propylester)	15,000	b
110-00-9	Furan	5,000	b
110-89-4	Piperidine	15,000	b
123-73-9	Crotonaldehyde, (E)- (2-Butenal, (E)-)	20,000	b
126-98-7	Methacrylonitrile (2-Propenenitrile, 2-methyl-)	10,000	b
151-56-4	Ethyleneimine (Aziridine)	10,000	b
302-01-2	Hydrazine	15,000	b
353-42-4	Boron trifluoride compound with methyl ether (1:1) (Boron, trifluoro(oxybis(methane))-, T-4-	15,000	b
506-77-4	Cyanogen chloride	10,000	c
509-14-8	Tetranitromethane (Methane, tetranitro-)	10,000	b
542-88-1	Chloromethyl ether (Methane, oxybis(chloro-)	1,000	b
556-64-9	Methyl thiocyanate (Thiocyanic acid, methyl ester)	20,000	b
584-84-9	Toluene 2,4-diisocyanate (Benzene, 2,4-diisocyanato-1-methyl-)[1]	10,000	a
594-42-3	Perchloromethylmercaptan (Methanesulfenyl chloride, trichloro-)	10,000	b
624-83-9	Methyl isocyanate (Methane, isocyanato-)	10,000	a, b
814-68-6	Acrylyl chloride (2-Propenoyl chloride)	5,000	b
4170-30-3	Crotonaldehyde (2-Butenal)	20,000	b
7446-09-5	Sulfur dioxide (anhydrous)	5,000	a, b
7446-11-9	Sulfur trioxide	10,000	a, b
7550-45-0	Titanium tetrachloride (Titanium chloride (TiCl4) (T-4)-)	2,500	b
7637-07-2	Boron trifluoride (Borane, trifluoro-)	5,000	b
7647-01-0	Hydrochloric acid (conc 30% or greater)	15,000	d
7647-01-0	Hydrogen chloride (anhydrous) (Hydrochloric acid)	5,000	a
7664-39-3	Hydrogen fluoride/Hydrofluoric acid (conc 50% or greater) (Hydrofluoric acid)	1,000	a, b
7664-41-7	Ammonia (anhydrous)	10,000	a, b
7664-41-7	Ammonia (conc 20% or greater)	20,000	a, b
7697-37-2	Nitric acid (conc 80% or greater)	15,000	b
7719-12-2	Phosphorus trichloride (Phosphorous trichloride)	15,000	b
7726-95-6	Bromine	10,000	a, b
7782-41-4	Fluorine	1,000	b
7782-50-5	Chlorine	2,500	a, b
7783-06-4	Hydrogen sulfide	10,000	a, b
7783-07-5	Hydrogen selenide	500	b

Appendix 7 - EPA Regulated Substances

CAS No.	Chemical name	Threshold quantity (lbs)	Basis for listing
7783-60-0	Sulfur tetrafluoride (Sulfur fluoride (SF4), (T-4)-)	2,500	b
7784-34-1	Arsenous trichloride	15,000	b
7784-42-1	Arsine	1,000	b
7803-51-2	Phosphine	5,000	b
8014-95-7	Oleum (Fuming Sulfuric acid) (Sulfuric acid, mixture with sulfur trioxide)[1]	10,000	e
10025-87-3	Phosphorus oxychloride (Phosphoryl chloride)	5,000	b
10049-04-4	Chlorine dioxide (Chlorine oxide (ClO$_2$))	1,000	c
10102-43-9	Nitric oxide (Nitrogen oxide (NO))	10,000	b
10294-34-5	Boron trichloride (Borane, trichloro-)	5,000	b
13463-39-3	Nickel carbonyl	1,000	b
13463-40-6	Iron, pentacarbonyl- (Iron carbonyl (Fe(CO)$_5$), (TB-5-11)-)	2,500	b
19287-45-7	Diborane	2,500	b
26471-62-5	Toluene diisocyanate (unspecified isomer) (Benzene, 1,3-diisocyanatomethyl-1)[1]	10,000	a

[1] The mixture exemption in 68.115(b)(1) does not apply to the substance.

Note: Basis for Listing:
a Mandated for listing by Congress.
b On EHS list, vapor pressure 10 mmHg or greater.
c Toxic gas.
e Toxicity of sulfur trioxide and sulfuric acid, potential to release sulfur trioxide, and history of accidents.

302 OSHA and EPA PSM Requirements

Table 3 to 68.130. - List of Regulated Flammable Substances and Threshold Quantities for Accidental Release Prevention
(Alphabetical Order--63 Substances)

Chemical name	CAS No.	Threshold quantity (lbs)	Basis for listing
Acetaldehyde	75-07-0	10,000	g
Acetylene (Ethyne)	74-86-2	10,000	f
Bromotrifluorethylene (Ethene, bromotrifluoro-)	598-73-2	10,000	f
1,3-Butadiene	106-99-0	10,000	f
Butane	106-97-8	10,000	f
1-Butene	106-98-9	10,000	f
2-Butene	107-01-7	10,000	f
Butene	25167-67-3	10,000	f
2-Butene-cis	590-18-1	10,000	f
2-Butene-trans (2-Butene, (E))	624-64-6	10,000	f
Carbon oxysulfide (Carbon oxide sulfide (COS))	463-58-1	10,000	f
Chlorine monoxide (Chlorine oxide)	7791-21-1	10,000	f
2-Chloropropylene (1-Propene, 2-chloro-)	557-98-2	10,000	g
1-Chloropropylene (1-Propene, 1-chloro-)	590-21-6	10,000	g
Cyanogen (Ethanedinitrile)	460-19-5	10,000	f
Cyclopropane	75-19-4	10,000	f
Dichlorosilane (Silane, dichloro-)	4109-96-0	10,000	f
Difluoroethane (Ethane, 1,1-difluoro-)	75-37-6	10,000	f
Dimethylamine (Methanamine, N-methyl-)	124-40-3	10,000	f
2,2-Dimethylpropane (Propane, 2,2-dimethyl-)	463-82-1	10,000	f
Ethane	74-84-0	10,000	f
Ethyl acetylene (1-Butyne)	107-00-6	10,000	f
Ethylamine (Ethanamine)	75-04-7	10,000	f
Ethyl chloride (Ethane, chloro-)	75-00-3	10,000	f
Ethylene (Ethene)	74-85-1	10,000	f
Ethyl ether (Ethane, 1,1'-oxybis-)	60-29-7	10,000	g
Ethyl mercaptan (Ethanethiol)	75-08-1	10,000	g
Ethyl nitrite (Nitrous acid, ethyl ester)	109-95-5	10,000	f
Hydrogen	1333-74-0	10,000	f
Isobutane (Propane, 2-methyl)	75-28-5	10,000	f
Isopentane (Butane, 2-methyl-)	78-78-4	10,000	g
Isoprene (1,3-Butadinene, 2-methyl-)	78-79-5	10,000	g
Isopropylamine (2-Propanamine)	75-31-0	10,000	g
Isopropyl chloride (Propane, 2-chloro-)	75-29-6	10,000	g
Methane	74-82-8	10,000	f
Methylamine (Methanamine)	74-89-5	10,000	f
3-Methyl-1-butene	563-45-1	10,000	f
2-Methyl-1-butene	563-46-2	10,000	g
Methyl ether (Methane, oxybis-)	115-10-6	10,000	f

Appendix 7 - EPA Regulated Substances

Chemical name	CAS No.	Threshold quantity (lbs)	Basis for listing
Methyl formate (Formic acid, methyl ester)	107-31-3	10,000	g
2-Methylpropene (1-Propene, 2-methyl-)	115-11-7	10,000	f
1,3-Pentadinene	504-60-9	10,000	f
Pentane	109-66-0	10,000	g
1-Pentene	109-67-1	10,000	g
2-Pentene, (E)-	646-04-8	10,000	g
2-Pentene, (Z)-	627-20-3	10,000	g
Propadiene (1,2-Propadiene)	463-49-0	10,000	f
Propane	74-98-6	10,000	f
Propylene (1-Propene)	115-07-1	10,000	f
Propyne (1-Propyne)	74-99-7	10,000	f
Silane	7803-62-5	10,000	f
Tetrafluoroethylene (Ethene, tetrafluoro-)	116-14-3	10,000	f
Tetramethylsilane (Silane, tetramethyl-)	75-76-3	10,000	g
Trichlorosilane (Silane, trichloro-)	10025-78-2	10,000	g
Trifluorochloroethylene (Ethene, chlorotrifluoro-)	79-38-9	10,000	f
Trimethylamine (Methanamine, N,N-dimethyl-)	75-50-3	10,000	f
Vinyl acetylene (1-Buten-3-yne)	689-97-4	10,000	f
Vinyl chloride (Ethene, chloro-)	75-01-4	10,000	a, f
Vinyl ethyl ether (Ethene, ethoxy-)	109-92-2	10,000	g
Vinyl fluoride (Ethene, fluoro-)	75-02-5	10,000	f
Vinylidene chloride (Ethene, 1,1-dichloro-)	75-35-4	10,000	g
Vinylidene fluoride (Ethene, 1,1-difluoro-)	75-38-7	10,000	f
Vinyl methyl ether (Ethene, methoxy-)	107-25-5	10,000	f

Note: Basis for Listing:
a Mandated for listing by Congress.
f Flammable gas.
g Volatile flammable liquid.

304 OSHA and EPA PSM Requirements

Table 4 to 68.130. - List of Regulated Flammable Substances and Threshold Quantities for Accidental Release Prevention
(CAS Number Order--63 Substances)

CAS No.	Chemical name	CAS No.	Threshold quantity (lbs)	Basis for listing
60-29-7	Ethyl ether (Ethane, 1,1'-oxybis-)	60-29-7	10,000	g
74-82-8	Methane	74-82-8	10,000	f
74-84-0	Ethane	74-84-0	10,000	f
74-85-1	Ethylene (Ethene)	74-85-1	10,000	f
74-86-2	Acetylene (Ethyne)	74-86-2	10,000	f
74-89-5	Methylamine (Methanamine)	74-89-5	10,000	f
74-98-6	Propane	74-98-6	10,000	f
74-99-7	Propyne (1-Propyne)	74-99-7	10,000	f
75-00-3	Ethyl chloride (Ethane, chloro-)	75-00-3	10,000	f
75-01-4	Vinyl chloride (Ethene, chloro-)	75-01-4	10,000	a, f
75-02-5	Vinyl fluoride (Ethene, fluoro-)	75-02-5	10,000	f
75-04-7	Ethylamine (Ethanamine)	75-04-7	10,000	f
75-07-0	Acetaldehyde	75-07-0	10,000	g
75-08-1	Ethyl mercaptan (Ethanethiol)	75-08-1	10,000	g
75-19-4	Cyclopropane	75-19-4	10,000	f
75-28-5	Isobutane (Propane, 2-methyl)	75-28-5	10,000	f
75-29-6	Isopropyl chloride (Propane, 2-chloro-)	75-29-6	10,000	g
75-31-0	Isopropylamine (2-Propanamine)	75-31-0	10,000	g
75-35-4	Vinylidene chloride (Ethene, 1,1-dichloro-)	75-35-4	10,000	g
75-37-6	Difluoroethane (Ethane, 1,1-difluoro-)	75-37-6	10,000	f
75-38-7	Vinylidene fluoride (Ethene, 1,1-difluoro-)	75-38-7	10,000	f
75-50-3	Trimethylamine (Methanamine, N, N-dimethyl-)	75-50-3	10,000	f
75-76-3	Tetramethylsilane (Silane, tetramethyl-)	75-76-3	10,000	g
78-78-4	Isopentane (Butane, 2-methyl-)	78-78-4	10,000	g
78-79-5	Isoprene (1,3,-Butadiene, 2-methyl-)	78-79-5	10,000	g
79-38-9	Trifluorochloroethylene (Ethene, chlorotrifluoro-)	79-38-9	10,000	f
106-97-8	Butane	106-97-8	10,000	f
106-98-9	1-Butene	106-98-9	10,000	f

Appendix 7 - EPA Regulated Substances 305

CAS No.	Chemical name	CAS No.	Threshold quantity (lbs)	Basis for listing
196-99-0	1,3-Butadiene	106-99-0	10,000	f
107-00-6	Ethyl acetylene (1-Butyne)	107-00-6	10,000	f
107-01-7	2-Butene	107-01-7	10,000	f
107-25-5	Vinyl methyl ether (Ethene, methoxy-)	107-25-5	10,000	f
107-31-3	Methyl formate (Formic acid, methyl ester)	107-31-3	10,000	g
109-66-0	Pentane	109-66-0	10,000	g
109-67-1	1-Pentene	109-67-1	10,000	g
109-92-2	Vinyl ethyl ether (Ethene, ethoxy-)	109-92-2	10,000	g
109-95-5	Ethyl nitrite (Nitrous acid, ethyl ester)	109-95-5	10,000	f
115-07-1	Propylene (1-Propene)	115-07-1	10,000	f
115-10-6	Methyl ether (Methane, oxybis-)	115-10-6	10,000	f
115-11-7	2-Methylpropene (1-Propene, 2-methyl-)	115-11-7	10,000	f
116-14-3	Tetrafluoroethylene (Ethene, tetrafluoro-)	116-14-3	10,000	f
124-40-3	Dimethylamine (Methanamine, N-methyl-)	124-40-3	10,000	f
460-19-5	Cyanogen (Ethanedinitrile)	460-19-5	10,000	f
463-49-0	Propadiene (1,2-Propadiene)	463-49-0	10,000	f
463-58-1	Carbon oxysulfide (Carbon oxide sulfide (COS))	463-58-1	10,000	f
463-82-1	2,2-Dimethylpropane (Propane, 2,2-dimethyl-)	463-82-1	10,000	f
504-60-9	1,3-Pentadiene	504-60-9	10,000	f
557-98-2	2-Chloropropylene (1-Propene, 2-chloro-)	557-98-2	10,000	g
563-45-1	3-Methyl-1-butene	563-45-1	10,000	f
563-46-2	2-Methyl-1-butene	563-46-2	10,000	g
590-18-1	2-Butene-cis	590-18-1	10,000	f
590-21-6	1-Chloropropylene (1-Propene, 1-chloro-)	590-21-6	10,000	g
598-73-2	Bromotrifluorethylene (Ethene, bromotrifluoro-)	598-73-2	10,000	f
624-64-6	2-Butene-trans (2-Butene, (E))	624-64-6	10,000	f
627-20-3	2-Pentene, (Z)-	627-20-3	10,000	g
646-04-8	2-Pentene, (E)-	646-04-8	10,000	g
689-97-4	Vinyl acetylene (1-Buten-3-yne)	689-97-4	10,000	f
1333-74-0	Hydrogen	1333-74-0	10,000	f
4109-96-0	Dichlorosilane (Silane, dichloro-)	4109-96-0	10,000	f

306 OSHA and EPA PSM Requirements

CAS No.	Chemical name	CAS No.	Threshold quantity (lbs)	Basis for listing
7791-21-1	Chlorine monoxide (Chlorine oxide)	7791-21-1	10,000	f
7803-62-5	Silane	7803-62-5	10,000	f
10025-78-2	Trichlorosilane (Silane,trichloro-)	10025-78-2	10,000	g
25167-67-3	Butene	25167-67-3	10,000	f

Note: Basis for Listing:
a Mandated for listing by Congress.
f Flammable gas.
g Volatile flammable liquid.

Appendix 8

Accidental Release Prevention Provisions (Proposed)

PART 68 - ACCIDENTAL RELEASE PREVENTION PROVISIONS

Subpart A - General Provisions
68.1 Scope.
68.3 Definitions.
68.5 Threshold Determination (Reserved).

Subpart B - Risk Management Plan Requirements
68.10 Applicability.
68.12 Registration.
68.15 Hazard assessment.
68.20 Prevention program purpose.
68.22 Prevention program - management system.
68.24 Prevention program - process hazard analysis.
68.26 Prevention program - process safety information.
68.28 Prevention program - standard operating procedures.
68.30 Prevention program - training.
68.32 Prevention program - maintenance.
68.34 Prevention program - pre-startup review.
68.36 Prevention program - management of change.
68.38 Prevention program - safety audits.
68.40 Prevention program - accident investigation.
68.45 Emergency response program.
68.50 Risk management plan.
68.55 Recordkeeping requirements.
68.60 Audits.

[Authority: 42 U.S.C. 7412(r) and 7601(a)(1).]

308 OSHA and EPA PSM Requirements

Subpart A - General Provisions

68.1 Scope.
This part sets forth requirements for chemical accident prevention steps that must be taken by the owner or operator of stationary sources.

68.3 Definitions.
As used in this part, all terms not defined shall have the meaning given to them by the Clean Air Act (42 U.S.C. 7401 et seq.).

Act means the Clean Air Act as amended (42 U.S.C. 7401 et seq.).

Administrator means the administrator of the U.S. Environmental Protection Agency.

Analysis of offsite consequences means a qualitative or quantitative analysis of a range of accidental releases, including worst-case releases, to determine offsite effects including potential exposures of affected populations.

Mitigation system means specific equipment, substances or personnel designed or deployed to mitigate an accidental release; examples of mitigation systems include water curtain sprays, foam suppression systems, and emergency response teams.

Offsite means areas beyond the property boundary of the stationary source or areas within the property boundary to which the public has routine and unrestricted access.

Owner or operator means any person who owns, leases, operates, or controls a stationary source.

RMP means the risk management plan required under section 68.50.

SIC means Standard Industrial Classification.

Significant accidental release means any accidental release of a regulated substance that has caused or has the potential to cause offsite consequences such as death, injury, or adverse effects to human health or the environment or to cause the public to shelter-in-place or be evacuated to avoid such consequences.

Worst-case release means the loss of all of the regulated substance from the process in an accidental release that leads to the worst offsite consequences.

68.5 Threshold determination. (Reserved.)

Subpart B - Risk Management Program Requirements

68.10 Applicability.

(a) The requirements in this subpart apply to all stationary sources that, after (three years from the date of final rule publication) have a regulated substance present in a process in more than a threshold quantity as determined under 68.5.

(b) Stationary sources covered by this subpart shall comply with 68.12 through 68.60 no later than (three years after the date of final rule publication) or within three years after the date on which a regulated substance first becomes present in a process in more than a threshold quantity.

68.12 Registration.

(a) By (three years after the publication date of the final rule), or within three years of the date on which a stationary source becomes subject to this subpart, the owner or operator of each stationary source covered by this part shall register with the Administrator.

(b) The registration shall include the following:

(1) The name of the stationary source, its street address, its mailing address, and telephone number;

(2) The names and CAS numbers of all regulated substances that are present at the stationary source in greater than the threshold quantities, and the maximum amount present in a process at any one time (in ranges);

(3) For each regulated substance, the four-digit SIC code(s) that apply to the use of the substance at the stationary source;

(4) The Dun and Bradstreet number of the stationary source;

(5) The name of a contact person; and

(6) The following certification signed by the owner or operator: "The undersigned certifies that, to the best of my knowledge, information, and belief formed after reasonable inquiry, the information submitted is true, accurate, and complete. I certify that I prepared or caused to be prepared a risk management plan that complies with 40 CFR 68.50 (and, when applicable:

"and the provisions of 40 CFR 68.60") and that I submitted or caused to be submitted copies of the risk management plan to each of the entities listed in 40 CFR 68.50(a). (Signature)."

(c) If at any time after the submission of the registration, information in the registration is no longer accurate, the owner or operator shall submit an amended notice within 60 days to the Administrator and implementing agency. After a final determination of necessary revisions under 68.60(f), the owner or operator shall register the revised risk management plan by the date required in 68.60(g).

68.15 Hazard assessment.

(a) The purpose of the hazard assessment is to evaluate the impact of significant accidental releases on the public health and environment and to develop a history of such releases.

(b) Hazard assessments shall be conducted for each regulated substance present at the stationary source above the threshold quantity. For each regulated substance, the hazard assessment shall include the following steps:

(1) Determine a worst-case release scenario for the regulated substance at the stationary source;

(2) Identify other more likely significant accidental releases for each process where the regulated substance is present above the threshold quantity, including processes where the substance is manufactured, processed, or used, and where the regulated substance is stored, loaded, or unloaded;

(3) Analyze the offsite consequences of the worst-case release scenario and the other more likely significant accidental release scenarios identified in 68.15(b)(2); and

(4) Develop a history of accidental releases of the regulated substance.

(c) To determine a worst-case release scenario, the owner or operator shall examine each process handling each regulated substance and assume that all of the regulated substance in the process is instantaneously released and all mitigation systems fail to minimize the consequences of the release.

(d) The owner or operator shall determine other more likely significant accidental releases such as but not limited to:

(1) Transfer hose failure, excess flow valve or emergency shutoff failure and subsequent loss of piping and shipping container contents (truck or rail);

(2) Process piping failure and loss of contents from both directions from the break; and

(3) Reactor or other process vessel failure where the contents are at temperatures and pressures above ambient

conditions. In these situations, passive mitigation systems are assumed to work to minimize the consequences of the release.

(e) For each regulated substance, the offsite consequences of the worst case or more likely significant accidental release scenarios shall be analyzed as follows:

(1) The rate and quantity of substance lost to the air and the duration of the event;

(2) The distance, in all directions, at which exposure to the substance or damage to offsite property or the environment from the release could occur using both worst-case meteorological conditions (i.e., F stability and 1.5 m/sec wind speed) and meteorological conditions most often occurring at the stationary source;

(3) Populations within these distances that could be exposed to the vapor cloud, pressure wave, or debris, depending on wind direction and meteorological conditions; and

(4) Environmental damage that could be expected within these distances, including consideration of sensitive ecosystems, migration routes, vulnerable natural areas, and critical habitats for threatened or endangered species.

(f) The owner or operator shall prepare a five-year history of significant accidental releases and releases with potential for offsite consequences for each regulated substance handled at the stationary source. The history shall list the release date, time, substance and quantity released, the duration of the release, the concentration of the substance released, and any offsite consequences such as deaths, injuries, hospitalizations, medical treatments, evacuations, sheltering in-place, and major off-site environmental impacts such as soil, groundwater, or drinking water contamination, fish kills, and vegetation damage.

(g) The hazard assessment shall be reviewed and updated at least once every five years. If changes in process, management, or any other relevant aspect of the stationary source or its surroundings (e.g. new housing developments or improved emergency response services) might reasonably be expected to make the results of the hazard assessment inaccurate (i.e., if either the worst-case release scenario or the estimate of offsite effects might reasonably be expected to change), the owner or operator shall complete a new or revised hazard assessment within 60 days of such change.

(h) The owner or operator shall maintain the following records documenting the hazard assessment and analysis of offsite consequences:

(1) A description of the worst-case scenario;

(2) A description of the other more likely significant accidental release scenarios identified in 68.15(b)(2), assumptions used, analyses or worksheets used to derive the accident scenarios, and the rationale for selection of specific scenarios; and

(3) Documentation for how the offsite consequences for each scenario were determined including:

(i) Estimated quantity of substance released, rate of release, and duration of the release;

(ii) Meteorological data used for typical conditions at the stationary source;

(iii) For toxic substances, the concentration used to determine the level of exposure and the data used for that concentration;

(iv) Calculations for determination of the distances downwind to the acute toxicity concentration; and

(v) Data used for estimation of the populations exposed or area damaged.

(i) A summary of the information required under paragraph (h) of this section and a table showing the data for the five-year accident history under paragraph (f) of this section shall be included in the RMP required under 68.50.

68.20 Prevention program purpose.

The owner or operator of a stationary source having one or more regulated substance above the threshold quantity shall develop and implement an integrated management system to evaluate the hazards present at the stationary source and to find the best ways to control these hazards. The prevention program includes ten required elements that must be tailored to suit the degree of hazards present at the stationary source and the degree of complexity of the stationary source's operations and that should work together under management control to ensure safe operations.

68.22 Prevention program - management system.

(a) The owner or operator of the stationary source shall develop a management system to oversee the implementation of the risk management program elements. The purpose of the management system is to ensure that the elements of the risk management program are integrated and implemented on an ongoing basis and that the responsibility for the overall program and for each element is clear.

(b) As part of the management system, the owner or operator shall identify a single person or position that has the overall responsibility for the development, implementation, and integration of the risk management program requirements.

(c) When responsibility for implementing individual requirements of the risk management program is assigned to persons other than the person designated under paragraph (b) of this section, the names or positions of these people shall be documented and the lines of authority defined through an organization chart or similar document.

68.24 Prevention program - process hazard analysis.

(a) The purpose of the process hazard analysis (hazard evaluation) is to examine, in a systematic, step-by-step way, the equipment, systems, and procedures for handling regulated substances and to identify the mishaps that could occur, analyze the likelihood that mishaps will occur, evaluate the consequences of these mishaps, and analyze the likelihood that safety systems, mitigation systems, and emergency alarms will function properly to eliminate or reduce the consequences of a mishap. A thorough process hazard analysis is the foundation for the remaining elements of the prevention program.

(b) The owner or operator shall perform an initial process hazard analysis on processes covered by this part. The process hazard analysis shall be appropriate to the complexity of the process and shall identify, evaluate, and control the hazards involved in the process. The owner or operator shall determine and document the priority order for conducting process hazard analyses based on a rationale which includes such considerations as the extent of process hazards, offsite consequences, age of the process, and operating history of the process. The process hazard analysis shall be completed no later than (three years after the date of final rule publication).

(c) Process hazard analyses completed after (Insert date 5 years before the effective date of the final rule) which meet the requirements of this section are acceptable as initial process hazard analyses. These process hazard analyses shall be updated and

314 OSHA and EPA PSM Requirements

revalidated, based on their completion date, in accordance with paragraph (h) of this section.

(d) The owner or operator shall use one or more of the following methodologies that are appropriate to determine and evaluate the hazards of the process being analyzed:
   (1) What-If;
   (2) Checklist;
   (3) What-If/Checklist;
   (4) Hazard and Operability Study (HAZOP);
   (5) Failure Mode and Effects Analysis (FMEA);
   (6) Fault Tree Analysis; or
   (7) An appropriate equivalent methodology.

(e) The process hazard analysis shall address:
   (1) The hazards of the process;
   (2) The identification of any previous incident which had a likely potential for significant offsite consequences;
   (3) Engineering and administrative controls applicable to the hazards and their interrelationships such as appropriate application of detection methodologies to provide early warning of releases. Acceptable detection methods might include process monitoring and control instrumentation with alarms, and detection hardware such as hydrocarbon sensors;
   (4) Consequences of failure of engineering and administrative controls;
   (5) Stationary source siting;
   (6) Human factors; and
   (7) A qualitative evaluation of a range of possible safety and health effects of failure of the controls on public health and the environment.

(f) The process hazard analysis shall be performed by a team with expertise in engineering and process operations, and the team shall include at least one employee who has experience and knowledge specific to the process being evaluated. Also, one member of the team must be knowledgeable in the specific process hazard analysis methodology being used.

(g) The owner or operator shall establish a system to promptly address the team's findings and recommendations; assure that the recommendations are resolved in a timely manner and that the resolution is documented; document what actions are to be taken; complete actions as soon as possible; develop a written schedule of when these actions are to be completed; and communicate the action to operating, maintenance, and other employees whose work assignments are in the process and who are affected by the recommendations or actions.

(h) At least every five (5) years after the completion of the initial process hazard analysis, the process hazard analysis shall be updated and revalidated by a team meeting the requirements in paragraph (f) of this section, to assure that the process hazard analysis is consistent with the current process.

(i) The owner or operator shall retain process hazard analyses and updates or revalidations for each process covered by this section, as well as the documented resolution of recommendations described in paragraph (g) of this section for the life of the process.

(j) Based on the findings and recommendations of the process hazard analysis, the owner or operator shall also investigate, evaluate, and document a plan for, or rationale for not, installing (if not already in place):

(1) Monitors, detectors, sensors, or alarms for early detection of accidental releases;

(2) Secondary containment or control devices such as, but not limited to, flares, scrubbers, quench, surge, or dump tanks, to capture releases; and

(3) Mitigation systems to reduce the downwind consequences of the release.

68.26 Prevention program - process safety information.

(a) The owner or operator shall complete a compilation of written process safety information before conducting any process hazard analysis required in 68.24. The compilation of written process safety information is to enable the owner or operator and the employees involved in operating the process to identify and understand the hazards posed by those processes involving regulated substances. This process safety information shall include information pertaining to the hazards of the regulated substances used or produced by the process, information pertaining to the technology of the process, and information pertaining to the equipment in the process.

(b) Information pertaining to hazards of the regulated substance in the process. This information shall consist of at least the following:

(1) Toxicity information;
(2) Permissible exposure limits;
(3) Physical data;

(4) Reactivity data;
(5) Corrosivity data;
(6) Thermal and chemical stability data; and
(7) Hazardous effects of inadvertent mixing of different materials that could foreseeably occur.

Note: MSDSs meeting the requirements of 29 CFR 1910.1200(g) may be used to comply with this requirement to the extent they contain the information required by this paragraph.

(c) Information pertaining to the technology of the process. Information concerning the technology of the process shall include at least the following:
(1) A block flow diagram or simplified process flow diagram;
(2) Process chemistry;
(3) Maximum intended inventory;
(4) Safe upper and lower limits for such items as temperatures, pressures, flows, or compositions; and,
(5) An evaluation of the consequences of deviations, including those affecting public health and the environment.

(d) Where the original technological information required by paragraph (c) of this section no longer exists, such information may be developed in conjunction with the process hazard analysis in sufficient detail to support the analysis.

(e) Information pertaining to the equipment in the process. Information pertaining to the equipment in the process shall include:
(1) Materials of construction;
(2) Piping and instrument diagrams (P&ID's);
(3) Electrical classification;
(4) Relief system design and design basis;
(5) Ventilation system design;
(6) Design codes and standards employed;
(7) Material and energy balances for processes built after the effective date of rule; and
(8) Safety systems (e.g., interlocks, detection, or suppression systems).

(f) The owner or operator shall document that equipment complies with recognized and generally accepted good engineering practices.

(g) For existing equipment designed and constructed in accordance with codes, standards, or practices that are no longer in general use, the owner or operator shall determine and document that the equipment is designed, maintained, inspected, tested, and operating in a safe manner.

68.28 Prevention program - standard operating procedures.

(a) The purpose of written standard operating procedures is to document the safe and proper way to operate and maintain processes and equipment, and to handle and store regulated substances at a stationary source. Procedures may be based on the process hazard analysis (hazard evaluation) information, successful past operating experience, manufacturers' recommendations, and applicable and appropriate codes and standards. The owner or operator shall consider the complexity of the process or stationary source to develop standard procedures.

(b) The owner or operator shall develop and implement written operating procedures that provide clear instructions for safely conducting activities involved in each covered process consistent with the process safety information and shall address at least the following elements:

(1) Steps for each operating phase:
(i) Initial startup;
(ii) Normal operations;
(iii) Temporary operations;
(iv) Emergency shutdown including the conditions under which emergency shutdown is required, and the assignment of shutdown responsibility to qualified operators to assure that emergency shutdown is executed in a safe and timely manner;
(v) Emergency operations;
(vi) Normal shutdown; and
(vii) Startup following a turnaround, or after an emergency shutdown.

(2) Operating limits:
(i) Consequences of deviation; and
(ii) Steps required to correct or avoid deviation.

(3) Safety and health considerations:
(i) Properties of, and hazards presented by, the substances used in the process;
(ii) Precautions necessary to prevent exposure, including engineering controls, administrative controls, and personal protective equipment;
(iii) Control measures to be taken if physical contact or airborne exposure occurs;

318 OSHA and EPA PSM Requirements

(iv) Quality control for raw materials and control of regulated substance inventory levels; and,

(v) Any special or unique hazards.

(4) Safety systems and their functions.

(c) Operating procedures shall be readily accessible to employees who work in or maintain a process.

(d) The operating procedures shall be reviewed as often as necessary to assure that they reflect current operating practice, including changes that result from changes in process chemicals, technology, and equipment, and changes to stationary sources. The owner or operator shall certify annually that these operating procedures are current and accurate.

(e) The owner or operator shall develop and implement safe work practices to provide for the control of hazards during operations involving lockout/tagout; confined space entry; opening process equipment or piping; and control over entrance into a stationary source by maintenance, contractor, laboratory, or other support personnel. These safe work practices shall apply to employees and contractor employees working on a facility.

68.30 Prevention program - training.

(a) The purpose of the training program is to ensure that each employee involved with regulated substances has learned and understands the procedures developed under 68.28. The owner or operator shall consider the complexity of the procedures, and the complexity of the process or stationary sources when developing training programs.

(b) Initial training.

(1) Each employee presently operating a process, and each employee before operating a newly assigned process shall be trained in an overview of the process and in the operating procedures as specified in 68.28. The training shall include emphasis on the specific safety and health standards, emergency operations including shutdown, and safe work practices applicable to the employee's job tasks.

(2) In lieu of initial training for those employees already involved in operating a process on the effective date of this rule, an owner or operator may certify in writing that the employee has the required knowledge, skills, and abilities to safely carry out the duties and responsibilities as specified in the operating procedures.

(c) Refresher training. Refresher training shall be provided at least every three years and more often if necessary to each employee involved in operating a covered process to assure that the employee understands and adheres to the current operating procedures in the process. The owner or operator, in consultation with the employees involved in operating the process, shall determine the appropriate frequency of refresher training.

(d) Training documentation. The owner or operator shall ascertain that each employee involved in operating a process has received and understood the training required by this section. The owner or operator shall prepare a record which contains the identity of the employee, the date of training, and the means used to verify that the employee understood the training.

(e) The owner or operator shall evaluate the effectiveness of the training program. A schedule for reviewing and revising the program shall be maintained at the stationary source.

68.32 Prevention program - maintenance (mechanical integrity).

(a) The purpose of the maintenance program is to determine and target the specific equipment that is identified through the process hazard analysis (hazard evaluation) or through operating experience as needing regular maintenance because failure of the equipment would lead to a significant accidental release. The owner or operator shall consider the complexity of the process or stationary source in developing the maintenance program.

(b) The owner or operator shall develop a list of equipment and controls the failure of which could result in a significant accidental release. As applicable, the equipment list shall include:

(1) Pressure vessels and storage tanks;

(2) Piping systems (including piping components such as valves);

(3) Relief and vent systems and devices;

(4) Emergency shutdown systems;

(5) Controls (including monitoring devices and sensors, alarms, and interlocks); and,

(6) Pumps.

(c) Written procedures. The owner or operator shall establish and implement written procedures to maintain the on-going integrity of process equipment.

(d) Training for process maintenance activities. The owner or operator shall train each employee involved in maintaining the on-going integrity of process equipment in an overview of that process and its hazards and in the procedures applicable to the employee's job tasks to assure that the employee can perform the job tasks in a safe manner and shall document the training as required in 68.30(d).

(e) Maintenance, inspections, and testing. For every item of equipment required to be listed under paragraph (b) of this section, the owner or operator shall develop a maintenance program to inspect, test, and maintain the equipment on an appropriate schedule to ensure that the equipment and controls continue to function according to specifications.

(1) Maintenance, inspections, and tests shall be performed on process equipment.

(2) Maintenance, inspection, and testing procedures shall follow recognized and generally accepted good engineering practices.

(3) The frequency of maintenance, inspections, and tests of process equipment shall be consistent with applicable manufacturers' recommendations and good engineering practices, and more frequently if determined to be necessary by prior operating experience.

(4) The owner or operator shall document each maintenance procedure, inspection, and test that has been performed on process equipment. The documentation shall identify the date of the maintenance/inspection/test; the name of the person who performed the maintenance/inspection/test; the serial number or other identifier of the equipment on which the maintenance, inspection, or test was performed; a description of the maintenance, inspection, and test that is performed; and the results of the inspection or test.

(f) Equipment deficiencies. The owner or operator shall correct deficiencies in equipment that are outside acceptable limits (defined in the process safety information in 68.26(c)(4) and (e)) before further use or in a safe and timely manner when necessary means are taken to assure safe operations.

(g) Quality assurance.

(1) In the construction of new plants and equipment, the owner or operator shall assure that equipment as it is fabricated is suitable for the process application for which they will be used.

(2) Appropriate checks and inspections shall be performed to assure that equipment is installed properly and consistent with design specifications and manufacturer's instructions.

(3) The owner or operator shall assure that maintenance materials, spare parts, and equipment are suitable for the process application for which they will be used.

68.34 Prevention program - pre-startup review.

(a) The purpose of the pre-startup review is to ensure that new or modified equipment is ready to properly and safely contain any new or previously handled regulated substance before that substance is introduced into the system. The owner or operator shall consider the complexity of the process or stationary source in developing the pre-startup review.

(b) The owner or operator shall perform a pre-startup safety review for new stationary sources and for modified stationary sources when the modification is significant enough to require a change in the process safety information.

(c) The pre-startup safety review shall confirm that prior to the introduction of regulated substances to a process:

(1) Construction and equipment is in accordance with design specifications;

(2) Safety, operating, maintenance, and emergency procedures are in place and are adequate;

(3) For new stationary sources, a process hazard analysis has been performed and recommendations have been resolved or implemented before startup; and modified stationary sources meet the requirements contained in management of change, 68.36; and

(4) Training of each employee involved in operating or maintaining a process has been completed and that employees are trained in any new emergency response procedures.

68.36 Prevention program - management of change.

(a) The purpose of a management of change program is to ensure that any alteration of equipment, procedures, substances, or processes are thoroughly analyzed to identify hazards, the consequences of failures, and impacts of the change on existing equipment, procedures, substances, and processes prior to implementation of the change.

(b) For process equipment, devices, or controls, replacement is not a change if the design, materials of construction, and parameters for flow, pressure, and temperature satisfy the design specifications of the device replaced.

(c) The owner or operator shall establish and implement written procedures to manage changes to process chemicals, technology, equipment, and procedures; and changes to stationary sources that affect a covered process.

(d) The procedures shall assure that the following considerations are addressed prior to any change:

(1) The technical basis for the proposed change;

(2) Impact of change on likelihood of a significant accidental release;

(3) Modifications to operating procedures;

(4) Necessary time period for the change; and,

(5) Authorization requirements for the proposed change.

(e) Employees involved in operating a process and maintenance and contract employees whose job tasks will be directly affected by a change in the process shall be informed of and trained in the change prior to the startup of the process or affected part of the process.

(f) If a change covered by this section results in a change in the process safety information required by 68.26, such information shall be updated accordingly.

(g) If a change covered by this section results in a change in the operating procedures or practices required by 68.28, such procedures or practices shall be updated accordingly.

68.38 Prevention program - safety audits.

(a) The safety audit consists of a periodic examination of the management systems and programs at the stationary source. The examination shall include a review of the documentation and implementation of the requirements of this subpart. The owner or operator shall consider the complexity of the process and of the process safety management program to develop the safety audit procedures, plans, and timing.

(b) The owners or operators shall certify that they have evaluated compliance with the provisions of this section at least every three years, to verify that the procedures and practices developed under this part are adequate and are being followed.

(c) The safety audit shall be conducted by at least one person knowledgeable in the process.

(d) A report of the findings of the audit shall be developed.

(e) The owner or operator shall promptly determine and document an appropriate response to each of the findings of the audit, and document that deficiencies have been corrected.

(f) The owner or operator shall retain the two most recent safety audit reports, as well as the documented actions in paragraph (e) of this section.

68.40 Prevention program - accident investigation.
(a) The purpose of the accident investigation is to learn the underlying causes of accidents to take steps to prevent them or similar accidental releases from recurring.

(b) The owner or operator shall establish and implement written procedures to investigate each significant accidental release.

(c) The owner or operator shall investigate each significant accidental release.

(d) An accident investigation shall be initiated as promptly as possible, but not later than 48 hours following the significant accidental release.

(e) An accident investigation team shall be established and consist of at least one person knowledgeable in the process involved, including a contract employee if the incident involved work of the contractor, and other persons with appropriate knowledge and experience to thoroughly investigate and analyze the significant accidental release.

(f) A report shall be prepared at the conclusion of the investigation which includes at a minimum:
 (1) Date of significant accidental release;
 (2) Date investigation began;
 (3) A description of the significant accidental release;
 (4) The factors that contributed to the significant accidental release, including its initiating event, and root cause or causes that may have increased the likelihood of the initiating event; and,
 (5) Any recommendations resulting from the investigation.

(g) The owner or operator shall establish a system to promptly address and resolve the accident report findings and recommendations. Resolutions and corrective actions shall be documented.

(h) The report shall be reviewed with all affected personnel whose job tasks are relevant to the significant accidental release findings including contract employees where applicable.

(i) Significant accidental release investigation reports shall be retained for five years.

324 OSHA and EPA PSM Requirements

68.45 Emergency response program.

(a) The purpose of the emergency response program is to prepare for response to and mitigation of accidental releases to limit the severity of such releases and their impact on the public health and environment.

(b) The owner or operator of a stationary source shall establish and implement an emergency response plan for responding to and mitigating accidental releases of regulated substances. The plan shall detail the steps all employees shall take in response to accidental releases and shall include:

(1) Evacuation routes or protective actions for employees not directly involved in responding to the release;

(2) Procedures for employees responding to the release, including protective equipment use;

(3) Descriptions of all response and mitigation technologies available at the stationary source; and

(4) Procedures for informing the public and emergency response agencies about releases.

(c) The owner or operator shall develop written procedures for the use of emergency response equipment and for its inspection, testing, and maintenance. The maintenance program for emergency response equipment shall be documented as required in 68.32(e)(4).

(d) For each regulated substance, the owner or operator shall document the proper first-aid and emergency medical treatment necessary to treat accidental human exposure.

(e) The owner or operator shall train all employees in relevant emergency response procedures and document the training as required under 68.30(d).

(f) The owner or operator shall conduct drills or exercises to test the plan and evaluate its effectiveness. Each drill or exercise shall be documented in writing and shall include findings of the drill or exercise that indicate aspects of the plan and procedures which need to be revised. Plans shall be revised based on the findings of the drills or exercises. The owner or operator shall document the response to each finding from a drill or exercise. For each finding requiring a change that is implemented, the schedule for implementing the change shall be documented.

(g) Each emergency response plan shall be coordinated with local emergency response plans developed under part 355 of this chapter by the local emergency planning committees and local emergency response agencies. Upon request of the local emergency

Appendix 8 - CARP Regulations 325

planning committee, the owner or operator shall promptly provide information to the local emergency planning committee necessary for developing and implementing the community emergency response plan.

(h) The owner and operator shall maintain a copy of the emergency response plan, including descriptions of all mitigation systems in place, at the stationary source.

68.50 Risk management plan.

(a) The owner or operator of a stationary source covered by this part shall submit a risk management plan (report) summarizing the key elements of its risk management program to the implementing agency and shall submit copies to the State Emergency Response Commission, the Local Emergency Planning Committee with jurisdiction for the area where the source is located, and the Chemical Safety and Hazard Investigation Board. Each report submitted by the stationary source shall address all regulated substances present at the stationary source in quantities above the threshold quantity.

(b) The report shall include a copy of the registration form, with updated information to ensure that the registration information is accurate.

(c) The report shall include, for each regulated substance, a summary of the hazard assessment and analysis of offsite consequences and accident history data required by 68.15(i).

(d) The report shall include, for the stationary source, a description of the major hazards (e.g., equipment failure, human error, natural phenomena, or other factors or a combination of such factors which could lead to a significant accidental release) identified through the process hazard analyses, a description of the consequences of a failure to control for each identified major hazard, a summary of all actions taken or planned to address these hazards, and how significant accidental releases are prevented or mitigated, or the consequences reduced by these actions. The purpose of the summary is to identify major hazards and provide an overview of the prevention program being implemented by the stationary source to prevent significant accidental releases. For each action taken to address a hazard, the report shall include the date on which the action was started (or is scheduled to start) and the actual or scheduled completion date. Where the same actions (e.g., training, certain controls, preventive maintenance programs, improved emergency response plan) address a number of hazards, the description may be organized by actions rather than hazards. If any requirement for the risk management program specified in this subpart is not covered in the summary of actions taken to address hazards, the report shall include a brief description of the stationary source's implementation of the requirement.

(e) The report shall include a summary of the stationary source's emergency response plan. The summary shall include:

(1) The procedures adopted to inform emergency response authorities and the public;

(2) The name or position of the point of contact between the stationary source and the public authorities;

(3) The dates of drills and exercises completed and planned and the results of completed drills; and

(4) A description of coordination with the local emergency planning committee.

(f) The report shall include a description of the management system developed to implement and coordinate the elements of the hazard assessment, prevention program, and emergency response program at the stationary source. The description shall define the person or position at the stationary source that is responsible for the overall implementation and coordination of the risk management program requirements. Where regulated substances are present above their threshold quantities at several locations at the stationary source or where responsibility for implementing individual requirements is delegated to separate groups at the stationary source, an organization chart shall be included to describe the lines of responsibility.

(g) The report shall include a certification by the owner or operator that, to the best of the signer's knowledge, information, and belief formed after reasonable inquiry, the information submitted is true, accurate, and complete.

(h) The report shall be reviewed and updated at least every five years and resubmitted to the implementing agency and copies shall be submitted to the State Emergency Response Commission, the Local Emergency Planning Committee, and the Chemical Safety and Hazard Investigation Board. If a change such as the introduction of a new regulated substance or process occurs that requires a revised or updated hazard assessment or process hazard analysis, then the report shall be updated and resubmitted within six months of the introduction of the new process or substance.

(i) The report shall be available to the public under section 114(c) of the Clean Air Act.

68.55 Recordkeeping requirements.

(a) The owner or operator of a stationary source covered by this part shall develop and maintain at the stationary source, for five years, records supporting the implementation of the risk management program and the development of the risk management plan.

(b) For the process hazard analysis, safety audit, and accident investigation, the records required to be maintained under paragraph (a) of this section shall include management's response to each recommendation that is required to be made, addressed, and documented under 68.24(g), 68.38(e), 68.40(f), and 68.40(g). For implemented recommendations and recommendations to be implemented, the documentation shall include the date (or scheduled date) for starting implementation and the date (or scheduled date) for completion of the implementation. For each recommendation not implemented, the documentation shall include an explanation of the decision.

(c) For pre-startup reviews and management of change, the documentation shall include the findings of the review and any additional steps (including a description of the steps and the reasons they were implemented) that were taken prior to implementation of the startup or change.

(d) The owner or operator shall maintain copies of all standard operating, maintenance, management of change, emergency response, and accident investigation procedures required under this part.

68.60 Audits.

(a) In addition to inspections for the purpose of regulatory development and enforcement of the Act, the implementing agency shall periodically audit RMPs registered under 68.12 in order to review the adequacy of such RMPs and require revisions of RMPs when necessary to assure compliance with 68.50.

(b) Stationary sources shall be selected for audits based on any of the following criteria:

(1) Accident history of the stationary source;

(2) Accident history of other stationary sources in the same industry;

(3) Quantity of regulated substances present at the stationary source;

(4) Location of the stationary source and its proximity to the public and sensitive environments;

(5) The presence of specific regulated substances;
(6) The hazards identified in the RMP; or
(7) A plan providing for neutral, random oversight.

(c) The implementing agency shall have access to the stationary source, supporting documentation, and any area where an accidental release could occur.

(d) Based on the audit, the implementing agency may issue an owner or operator of a stationary source a written preliminary determination of necessary revisions to the source's RMP in order to assure that the RMP meets the criteria of 68.50 and reflects the purposes of subpart B of this part. This preliminary determination shall include an explanation for the basis for the revisions, reflecting industry standards and guidelines (such as AIChE/CCPS guidelines and ASME and API standards) to the extent that such standards and guidelines are applicable, and shall include a timetable for their implementation.

(e) Written response to a preliminary determination:

(1) The owner or operator shall respond in writing to a preliminary determination made in accordance with paragraph (d) of this section. The response shall state that the owner or operator will implement the revisions contained in the preliminary determination in accordance with the timetable included in the preliminary determination or shall state that the owner rejects the revisions in whole or in part. For each rejected revision, the owner or operator shall explain the basis for rejecting such revision. Such explanation may include substitute revisions.

(2) The written response under paragraph (e)(1) of this section shall be received by the implementing agency within 90 days of the issuance of the preliminary determination or a shorter period of time as the implementing agency specifies in the preliminary determination as necessary to protect human health and the environment. Prior to the written response being due and upon written request from the owner or operator, the implementing agency may provide in writing additional time for the response to be received.

(f) After providing the owner or operator an opportunity to respond under paragraph (e) of this section, the implementing agency may issue the owner or operator a written final determination of necessary revisions to the source's RMP. The final determination may adopt or modify the revisions contained in the preliminary determination under paragraph (d) of this section or may adopt the substitute revisions provided in the

Appendix 8 - CARP Regulations 329

response under paragraph (e) of this section. A final determination that adopts a revision rejected by the owner or operator shall include an explanation of the basis for the revision. A final determination that fails to adopt a substitute revision provided under paragraph (e) of this section shall include an explanation of the basis for finding such substitute revision unreasonable.

(g) Thirty (30) days after the issuance of a final determination under paragraph (f) of this section, the owner or operator shall be in violation of 68.12, 68.50(a), and 68.60 unless the owner or operator revises the RMP prepared under 68.50 as required by the final determination, submits copies of the revised RMP to the entities identified in 68.50(a), and registers the revised plan as provided in 68.12 (b) and (c).

(h) The public shall have access to the preliminary determinations, responses, and final determinations under this section.

(i) Nothing in this section shall preclude, limit, or interfere in any way with the authority of EPA or the state to exercise its enforcement, investigatory, and information gathering authorities concerning this part under the Clean Air Act.

Appendix 9

OSHA Compliance Directive CPL 2-2.45A

U.S. Department of Labor Assistant Secretary for
Occupational Safety and Health
Washington, D C 20210

OSHA Instruction CPL 2-2.45A

Directorate of Compliance Programs

Subject: 29 CFR 1910.119, Process Safety Management of Highly Hazardous Chemicals--Compliance Guidelines and Enforcement Procedures

A. <u>Purpose</u>. This instruction establishes uniform policies, procedures, standard clarifications, and compliance guidance for enforcement of the standard for Process Safety Management of Highly Hazardous Chemicals, 29 CFR 1910.119 ("PSM standard"), and amendments to the standard for Explosives and Blasting Agents, 29 CFR 1910.109.

B. <u>Scope</u>. This instruction applies OSHA-wide.

C. <u>References</u>.

1. 29 CFR 1910.119, Process Safety Management of Highly Hazardous Chemicals; Final Rule; February 24, 1992, <u>Federal Register</u> Vol. 57, No. 36, pp. 6356-6417.

2. OSHA Instruction CPL 2.45B, June 15, 1989, the Field Operations Manual (FOM).

3. OSHA Instruction STP 2.22A, CH-2, January 29, 1990, State Plan Policies and Procedures Manual.

4. OSHA Instruction CPL 2.94, July 22, 1991, OSHA Response to Significant Events of Potentially Catastrophic Consequence.

5. OSHA Instruction ADM 1-1.12B, December 29, 1989, Integrated Management Information System (IMIS) Forms Manual.

D. <u>Cancellation</u>. This instruction cancels:

1. OSHA Instruction CPL 2-2.45, September 6, 1988, Systems Safety Evaluation of Operations with Catastrophic Potential.

2. OSHA Notice CPL 2, March 9, 1992, Special Emphasis Program in Petrochemical Industries, Standard Industrial Classification (SIC) Codes 2821, 2869, and 2911.

Appendix 9 - OSHA Compliance Directive

OSHA Instruction CPL 2-2.45A

Directorate of Compliance Programs

E. <u>Action</u>. OSHA Regional Administrators and Area Directors shall ensure that all compliance and enforcement activities related to the PSM standard adhere to the guidelines of this instruction.

F. <u>Federal Program Change</u>. This instruction describes a Federal program change which affects State programs. Each Regional Administrator shall:

1. Ensure that a copy of this change is promptly forwarded to each State designee, using a format consistent with the Plan Change Two-Way Memorandum in Appendix P of OSHA Instruction STP 2.22A, CH-3.

2. Explain the technical content of this change to the State designees as requested.

3. Advise the State designees that, in order to ensure uniform enforcement of the Explosives and Blasting Agents Standard and the Process Safety Management of Highly Hazardous Chemicals Standard addressed by this instruction, State implementation of the procedures in this instruction, or comparable State procedures, must be carefully coordinated with OSHA.

4. Coordinate with the State to ensure appropriate staff training (as discussed at I.4. of this instruction), participation in the Program-Quality-Verification inspection scheduling process (as discussed at J.3.d. of this instruction), and to provide appropriate technical assistance.

5. Ensure that State designees are asked to acknowledge receipt of this Federal program change in writing to the Regional Administrator as soon as the State's intention is known, but not later than 70 calendar days after the date of issuance (10 days for mailing and 60 days for response). This acknowledgment must include the State's intention to follow OSHA's policies and procedures described in this instruction, or a description of the State's alternative policy and/or procedure which is "at least as effective" as the Federal policy and/or procedure.

6. Ensure that the State designees submit a plan supplement, in accordance with OSHA Instruction STP 2.22A, CH-2, as appropriate, following the established schedule that is agreed upon by the State and the

331

332 OSHA and EPA PSM Requirements

OSHA Instruction CPL 2-2.45A

Directorate of Compliance Programs

Regional Administrator to submit non-Field Operations Manual/Technical Manual Federal Program Changes.

    a. If the State intends to follow the revised inspection procedures described in this instruction, the State must submit either a revised version of this instruction, adapted as appropriate to reference State law, regulations and administrative structure, or a cover sheet describing how references in this instruction correspond to the State's structure. The State's acknowledgment letter may fulfill the plan supplement requirement if the appropriate documentation is provided.

    b. If the State adopts an alternative to Federal enforcement inspection procedures, the State's plan supplement must identify and provide a rationale for all substantial differences from Federal procedures in order for OSHA to judge whether a different State procedure is as effective as the comparable procedure.

7. After Regional review of the State plan supplement and resolution of any comments thereon, forward the State submission to the National Office in accordance with established procedures. The Regional Administrator shall provide a judgment on the relative effectiveness of each substantial difference in the State plan change and an overall assessment thereon with a recommendation for approval or disapproval by the Assistant Secretary.

8. Review policies, instructions, and guidelines issued by the States to determine that this change has been communicated to State program personnel.

G. <u>Background</u>. On February 24, 1992, OSHA promulgated the Final Rule for Process Safety Management of Highly Hazardous Chemicals. This standard originally became effective on May 26, 1992. An administrative stay delayed the effective date of paragraphs (f), (h), (j), and (l) until August 26, 1992. That stay has expired and the stayed provisions are now fully effective.

    1. In recent years, a number of catastrophic accidents in the chemical industry have drawn attention to the safety of processes involving highly hazardous chemicals. OSHA has determined that employees have

Appendix 9 - OSHA Compliance Directive

OSHA Instruction CPL 2-2.45A

Directorate of Compliance Programs

been and continue to be exposed in their workplaces to the hazards of releases of highly hazardous chemicals which may be toxic, reactive, flammable, or explosive.

2. The requirements of the PSM standard are intended to eliminate or mitigate the consequences of such releases. The standard emphasizes the application of management controls when addressing the risks associated with handling or working near hazardous chemicals.

3. In addition, the PSM standard has been developed in fulfillment of OSHA's obligation under the Clean Air Act Amendments (CAAA) of 1990, section 304(a). The final rule is consistent with the mandate of the CAAA.

4. It is anticipated that joint inspection activities related to the PSM standard will arise between OSHA, the Environmental Protection Agency, and the Chemical Safety and Hazard Investigation Board, which was mandated by the CAAA.

H. Enforcement Activity Related to the PSM Standard--Types of Inspections. 29 CFR 1910.119 has broad applicability to potentially hazardous processes that may exist in a wide variety of industries. Accordingly, enforcement activities related to the PSM standard--either to determine if an employer is covered by the standard or to assess the employer's compliance with it--may take place in any of the inspection types described below. The following guidelines shall apply to PSM-related compliance activity:

1. Program-Quality-Verification (PQV) Inspections. The primary enforcement model for the PSM standard shall be the PQV inspection, as described at F. and L. of this instruction. Programmed PQV inspections shall be scheduled as described at J. of this instruction.

2. Other Programmed Inspections: Screening for PSM Coverage. In all programmed safety and health inspections in general industry, a determination shall be made as to whether the establishment is covered by the PSM standard.

   a. This determination shall follow the criteria presented at 29 CFR 1910.119(a), including appropriate reference to Appendix A of §1910.119. The determination may be made in conjunction with

334 OSHA and EPA PSM Requirements

OSHA Instruction CPL 2-2.45A

Directorate of Compliance Programs

an assessment of the employer's Hazard Communication program.

   b. If the establishment is found to be covered by the standard:

      (1) It shall be further determined if the establishment is included in the universe of affected establishments from which PQV inspections may be scheduled. (See J. of this instruction.)

      (2) The employer shall be provided:

         (a) Copies of the OSHA publications "Process Safety Management," OSHA Publication 3132, which also contains the full text of §1910.119; and "Process Safety Management--Guidelines for Compliance;" and

         (b) A letter notifying the employer that the subject establishment is covered by the PSM standard and may be inspected under the standard. The letter shall also emphasize the employer's obligation to comply with the standard. An example of such a letter is provided as Appendix F of this instruction.

   c. The Area Director shall ensure proper coding of the OSHA-1 (as described at Q. and Appendix H of this instruction) to identify the establishment as either known to be covered by the PSM standard or known not to be covered by the standard.

3. <u>Unprogrammed PSM-related Inspections</u>. In all unprogrammed inspection activity relating to the PSM standard, a determination shall be made as to whether the establishment is covered by 29 CFR 1910.119.

   a. If a formal complaint or referral relating to the PSM standard is received regarding any workplace classified in one of the SIC codes listed at Appendix C of this instruction, the complaint or referral item(s) shall be investigated and:

Appendix 9 - OSHA Compliance Directive 335

OSHA Instruction CPL 2-2.45A

Directorate of Compliance Programs

        (1) All programs required by the PSM standard shall be screened for obvious violations; and

        (2) A CSHO referral for a PQV inspection shall be considered if major deficiencies are indicated. This determination shall be documented in the case file.

    b. Investigations of formal, PSM-related complaints and referrals in establishments in all other SIC codes shall normally be limited to the complaint item(s) only, unless violations related to the complaint or referral items are found.

4. <u>Responses to Accidents and Catastrophes</u>. Responses to accidents and catastrophes involving PSM shall follow the guidelines contained in Chapter VIII of the FOM and--where appropriate--in OSHA Instruction CPL 2.94, "OSHA Response to Significant Events of Potentially Catastrophic Consequence," in addition to the guidelines of this instruction. If the workplace is classified in one of the SIC codes listed at Appendix C of this instruction, a PQV inspection shall be considered; the reasons for the determination shall be documented in the case file.

5. <u>All Other Inspections</u>. Normally, there shall be no PSM-related activity on any inspection other than those described at H.1. through H.4., above.

I. <u>Inspection Resources</u>. Appropriate levels of staff training and preparation are essential for compliance activities relating to the PSM standard. In particular, it is anticipated that PQV inspections will be highly resource-intensive; they will therefore require careful planning and coordination. The recommendations included as Appendix G of this instruction may be used as a guide for such planning.

    1. <u>PQV Team Leaders ("Level One")</u>. Only trained compliance safety and health officers (CSHOs) with experience in the chemical industry shall be assigned to lead a PQV inspection under this standard.

        a. As a minimum, this training must include the OSHA Training Institute's Course 330, "Safety and Health in the Chemical Processing Industries," and Course 340, "Hazard Analysis in the Chemical Processing Industries," or equivalent training

336 OSHA and EPA PSM Requirements

OSHA Instruction CPL 2-2.45A

Directorate of Compliance Programs

such as that offered by the National Institute of Standards and Technology.

NOTE: Due to a significant change in course content, completion of Course 330 prior to Fiscal Year 1991 does not meet this requirement for PQV team leaders.

    b. Team leaders must have prior experience in the chemical industry. This experience should include experience obtained from accident/explosion investigations in chemical or petrochemical plants, through previous chemical inspections involving process safety management evaluation, or through previous chemical industry employment.

2. **PQV Team Members ("Level Two")**. CSHOs may be assigned as PQV team members, or to conduct unprogrammed inspections in workplaces in the targeted SIC codes listed in Appendix C of this instruction, if they have 2 years of OSHA inspection experience or the equivalent and have completed Course 330, "Safety and Health in the Chemical Processing Industries" (including offerings of this course prior to Fiscal Year 1991) and Course 340, "Hazard Analysis in the Chemical Processing Industries."

3. **CSHOs With Less Training**. Complaint and other unprogrammed inspections pertaining to some sections of the standard may be conducted by CSHOs who do not have the training and experience described at I.1. or I.2., above, but who are experienced in evaluating other programmatic standards such as hazard communication and lockout/tagout and in evaluating respirator programs.

    a. The following sections of 29 CFR 1910.119 may be appropriately evaluated by such CSHOs:

- (c) Employee participation.
- (g) Training.
- (h) Contractors.
- (k) Hot work permits.
- (m) Incident investigation.

Appendix 9 - OSHA Compliance Directive 337

OSHA Instruction CPL 2-2.45A

Directorate of Compliance Programs

          (n) Emergency planning and response.

    b. Such CSHOs shall make full utilization of Technical Support resources at the Regional Office and National Office levels in arriving at decisions regarding compliance or noncompliance.

    c. Nevertheless, to the extent possible, Area Directors shall attempt to utilize CSHOs with experience and training in the chemical industry to perform such unprogrammed inspections.

4. **State Plan States.** Each State shall have one or more CSHOs trained to meet the requirements for PQV team leaders and an appropriate number of qualified team members. OSHA will provide technical assistance, as needed, through the Regional Office, Health Response Team, and the Office of Construction and Engineering.

J. **PQV Inspection Scheduling.** Due to the resource-intensive nature of inspections for compliance with the PSM standard, the Agency will be able to perform only a limited number of PQV inspections (as described at K. and L. of this instruction) each year. A special targeting and scheduling system is therefore necessary to maximize the effective use of inspection resources.

1. **Targeting.** OSHA wishes to make the most effective use of its limited resources, and therefore will use the factors listed below in determining the SIC codes to be inspected. OSHA will select the SICs that have experienced the greatest number of accidents/incidents as determined from these three sources:

    a. Published insurance industry reports of major accidents/incidents.

    b. IMIS data, including the OSHA-170 Investigation Summary File.

    c. EPA Accident Release Information Program (ARIP) data.

2. **Current Targeted SICs.** A list of targeted SIC codes based on current data, as described at J.1., is included as Appendix C of this instruction. This Appendix may be updated periodically.

338 OSHA and EPA PSM Requirements

OSHA Instruction CPL 2-2.45A

Directorate of Compliance Programs

3. <u>Scheduling</u>. PQV inspections shall be scheduled as follows:

   a. Using the list of SICs determined as described at J.1. above, the Office of Statistics shall annually create an initial list including all known establishments within each of the identified SICs for each Region. This list shall be organized by establishment, by establishment size, by corporate identity (as determined through a commercially available source), and by State.

   b. The Directorate of Compliance Programs shall forward the initial lists to the appropriate Regions.

   c. Within 30 days of receipt of the initial list, each Region shall select five candidates for a PQV inspection and shall forward the resulting list, together with documentation supporting the selections, to the Directorate of Compliance Programs. The Regions shall base their selection on such factors as:

      (1) Number of employees at the facility.

      (2) Age of the facility.

      (3) Known toxicity of chemicals used in the facility's processes.

      (4) Frequency of media reports of releases or other incidents at the facility.

      (5) Local EPA information.

      (6) Past OSHA history of the facility, including complaints received and/or followup inspections due.

      (7) Information from local/municipal fire departments.

      NOTE: The Regions need not make a determination on each of these factors for each establishment on their initial list; however, their selections for

Appendix 9 - OSHA Compliance Directive 339

OSHA Instruction CPL 2-2.45A

Directorate of Compliance Programs

        candidates shall be thoroughly documented.

d. Beginning with Fiscal Year 1994, Regional Administrators shall provide to each State designee a copy of the establishment list within their State. Each State shall nominate one establishment for a PQV inspection or provide an explanation of why a PQV inspection should not be scheduled in their State. Regional Administrators shall include these candidates with their regional submission (i.e., in addition to the five candidates submitted by each Region) to the Directorate of Compliance Programs, together with their assessment and recommendation as to whether the State's candidate should be included in the national selection and as to the State's degree of readiness to conduct the inspection independently.

e. Within 60 days of the receipt of the candidate lists from the Regions, the Directorate of Compliance Programs shall notify each Region of the final list of establishments from which PQV inspections are to be scheduled.

  (1) The selections for the Regional lists shall be made by the Directorate of Compliance Programs in coordination with the Office of Field Programs. The selections shall be based on:

    (a) Emphasis on a corporate approach, to give inspection priority to the maximum number of different corporations, rather than targeting multiple inspections in the same corporation; and

    (b) Regional resources and inspection goals; and

    (c) Overall Agency resources.

  (3) Inspection goals (actual numbers) are to be set in the annual Field Operations Program Plan between Regions and the Office of Field Programs.

340 OSHA and EPA PSM Requirements

OSHA Instruction CPL 2-2.45A

Directorate of Compliance Programs

(4) The number of establishments selected may vary from Region to Region, because OSHA plans to focus more PQV inspections in Regions with higher concentrations of high-hazard industries affected by the standard.

(5) The selections from the States' list shall be coordinated with the Office of State Programs. Regional Administrators shall include full discussion and coordination with the affected States.

4. <u>Deletion Criteria</u>. An establishment shall be deleted from the list if it:

   a. Has received a substantially complete systems safety inspection or PQV inspection within the current or the preceding 5 calendar years; or

   b. Is included in a corporate settlement agreement requiring appropriate management systems for process safety; or

   c. Is a VPP participant; or

   d. Is a corporate office/headquarters and is not engaged in actual production or physical research operations; or

   e. Is not covered because of exclusions in the PSM standard; or

   f. Has been the subject of a PSM-related inspection (complaint or referral) in the preceding year during which PSM programs were screened and a referral for a PQV inspection was not made.

   NOTE: Determination for deletion shall be made initially, to the extent possible, at the National Office level when the list is prepared; and/or subsequently, as necessary, at the Regional Office level based on local knowledge (e.g., recent inspections, Area Office screening, State Manufacturers' Guide).

5. <u>Local Emphasis Programs</u>. Some Regions may have relatively few establishments in the targeted SIC codes

Appendix 9 - OSHA Compliance Directive

OSHA Instruction CPL 2-2.45A
Directorate of Compliance Programs

listed in Appendix C. Regional Administrators therefore may propose a Local Emphasis Program to direct Regional PSM inspections resources to industry types that may not be covered by the targeted SIC codes. Such Local Emphasis Programs shall be submitted to the Directorate of Compliance Programs for approval in accordance with the FOM, Chapter II.

K. **Scope of PQV Inspection.** Comprehensive inspections under the PSM standard shall evaluate the procedures used by the employer and the process-related contract employers to manage the hazards associated with processes using highly hazardous chemicals. Normally, these inspections will embody a three-fold approach, which for reference is termed **Program-Quality-Verification (PQV)**.

   1. First, the employer's and the contract employers' **Program** for complying with each of the listed elements of the PSM standard shall be evaluated in accordance with the PSM Audit Guidelines contained in Appendix A of this instruction. (See also M. of this instruction.)

   2. Second, the **Quality** of the employer's and the contract employers' procedures shall be compared to acceptable industry practices as described in the standard to determine compliance.

   3. Third, **Verification** of the employer's and the contract employers' effective implementation of the program can be made through review of written programs and records of activity, interviews with employees at different levels, and observation of site conditions. The team leader shall select one or more processes as described at L.7. of this instruction to perform the verification portion of the inspection.

L. **PQV Inspection Procedures.** The procedures given in the FOM, Chapter III, shall be followed except as modified in the following sections:

   1. **Opening Conference.** Where appropriate, the facility safety and health director, Process Safety Manager, or other person capable of explaining the company's Process Safety Management Program shall be included in the opening conference.

342   OSHA and EPA PSM Requirements

OSHA Instruction CPL 2-2.45A

Directorate of Compliance Programs

    a. During the opening conference, CSHOs shall familiarize themselves with the establishment's emergency response procedures and emergency alarms.

    b. CSHOs shall also request that the management representative(s) provide them with a reasonably detailed overview of the chemical process at the facility, including block diagrams indicating chemicals and processes involved.

2. __PSM Overview__. Prior to beginning the walkaround inspection, the CSHOs shall request an explanation of the company's Process Safety Management Program including, at a minimum:

    a. How the elements of the standard are implemented;

    b. Personnel designated as responsible for implementation of the various elements of the standard; and

    c. A description of company records used to verify compliance with the standard. (See also Appendix E of this instruction.)

3. __Initial Walkaround__. After this familiarization, the inspection may begin with a brief walkaround inspection of those portions of the facility within the scope of the standard. Additional walkaround activity may be necessary after selection of the process unit(s). The purpose of the initial walkaround is to:

    a. Give CSHOs a basic overview of the facility operations;

    b. Allow CSHOs to observe potential hazards such as pipework in risk of impact, corroded or leaking equipment, unit or control room siting, and location of relief devices; and

    c. Solicit input from the employee representative concerning potential PSM program deficiencies.

4. __Personal Protective Equipment__. In addition to normal inspection protective equipment, CSHOs conducting these inspections shall be provided with flame retardant coveralls for protection from flash fires and with

Appendix 9 - OSHA Compliance Directive 343

OSHA Instruction CPL 2-2.45A

Directorate of Compliance Programs

NIOSH-approved emergency escape respirators for use during any emergency conditions.

    a. CSHOs shall wear flame-retardant coveralls in all areas of the plant where there is potential for flash fires and as may be required by company policy.

        NOTE: Clothing made of hazardous synthetic fabrics should not be worn underneath flame-retardant coveralls.

    b. CSHOs shall carry emergency escape respirators, when necessary, during the walkaround portion(s) of the inspection. CSHOs conducting these inspections shall have received proper training in the use of emergency escape respirators.

    c. CSHOs shall be provided with appropriate alert monitors (e.g., HCN, Cl_2) where such devices are necessary.

    d. CSHOs shall ensure that any still cameras and/or video cameras are intrinsically safe for use in the process areas being inspected.

5. <u>Documentation to be Requested--General and Process-Related</u>. At the conclusion of the opening conference, the CSHO shall request access to or copies of the documents listed at L.5.a. through L.5.m. below. Initially, to expedite the inspection process, only access to documents should be requested. During the inspection, as potential violations of the standard are observed, copies of the written documentation described below shall be requested to substantiate citations.

    a. OSHA 200 Logs for the past 3 years for both the employer and all process-related contractor employer(s).

    b. Employer's written plan of action regarding the implementation of employee participation.

    c. Written process safety information for the unit(s) selected (see L.7.), if available, such as flow diagrams, piping and instrumentation diagrams (P&ID's), and process narrative descriptions.

344 OSHA and EPA PSM Requirements

OSHA Instruction CPL 2-2.45A

Directorate of Compliance Programs

NOTE: The employer is required to compile process safety information on a schedule consistent with the employer's schedule for conducting the process hazard analyses (PHA).

d. Documented priority order and rationale for conducting process hazard analyses; copies of any process hazard analyses performed after May 25, 1987; team members; actions to promptly address findings; written schedules for actions to be completed; documentation of resolution of findings; documentation verifying communication to appropriate personnel; and 5-year revalidation of original PHA required by standard.

e. Written operating procedures for safely conducting activities in each selected unit; annual certification that operating procedures are current and accurate; written procedures describing safe work practices for potentially hazardous operations, including (but not limited to) lockout/tagout, confined space entry, lifting equipment over process lines, capping over ended valves, opening process equipment or piping, excavation; and control over entrance into a facility of maintenance, laboratory, or other support personnel.

f. Training records for initial and refresher training for all employees in the selected unit(s) whose duties involve operating a process; methods for determining the content of the training; methods for determining frequency of refresher training; certification of required knowledge, skills, and abilities to safely perform job for employees already involved in operating a process on May 26, 1992, who have not received initial training; and training material.

g. Pre-startup safety review for new facilities and for modified facilities when the modification is significant enough to require a change in the process safety information; documentation of employee training.

h. Written procedures and schedules to maintain the ongoing integrity of process equipment; the

Appendix 9 - OSHA Compliance Directive 345

OSHA Instruction CPL 2-2.45A

Directorate of Compliance Programs

  relevant portions of applicable manufacturers' instructions, codes, and standards; and inspection and tests performed on process equipment in the unit(s) selected.

 i. Hot work permit program and active permits issued for the unit(s) selected.

 j. Written procedures to manage change to process chemicals, technology, equipment and procedures; and changes to facilities that affect a covered process.

 k. Incident investigation reports for the unit(s) selected, resolutions and corrective actions.

 l. Written emergency action plan including procedures for handling small releases and evidence of compliance with 1910.120(a), (p), and (q), where applicable.

 m. The two most recent compliance audit reports, appropriate responses to each of the findings, and verifications that deficiencies have been corrected.

6. <u>Documentation to be Requested--Contractor-Related</u>. The following information relating to contractor compliance shall be requested:

 a. <u>Documentation from Employer</u>:

  (1) Information relating to contract employers' safety performance and programs;

  (2) Methods of informing contract employers of known potential hazards related to contractor's work and the process and applicable provisions of the emergency action plan;

  (3) Safe work practices to control the entrance, presence and exit of contract employers and contract employees in covered process areas;

  (4) Evaluation of contractor employer performance in fulfilling responsibilities required by the standard;

346 OSHA and EPA PSM Requirements

OSHA Instruction CPL 2-2.45A

Directorate of Compliance Programs

    (5) Contract employee injury and illness logs related to work in process areas; and

    (6) A list of unique hazards presented by contractors' work or hazards found in the workplace that have been reported to the employer.

  b. <u>Documentation from Contract Employer</u>:

    (1) Records showing employees receive training in and understand safe work practices related to the process on or near which they will be working to perform their jobs safely;

    (2) Known potential fire, explosion or toxic release hazards related to job, and applicable provisions of emergency action plan; and

    (3) A list of unique hazards presented by contractors' work or hazards found in the workplace that have been reported to the employer.

NOTE: The documentation described at L.5. and L.6.a. may also be required of the contract employer, depending on the scope of the contract employer's activities.

7. <u>Selection of Process(es)</u>. The team leader shall select one or more processes within which to evaluate compliance with the standard. This selection shall be based on the factors listed below, and shall be documented in the case file:

  a. Factors observed during the walkthrough;

  b. Incident reports and other history;

  c. Company priorities for or completed process hazard analyses (PHA);

  d. Age of the process unit;

  e. Nature and quantity of chemicals involved;

  f. Employee representative input;

Appendix 9 - OSHA Compliance Directive

OSHA Instruction CPL 2-2.45A

Directorate of Compliance Programs

      g. Current hot work, equipment replacement, or other maintenance activities; and

      h. Number of employees present.

M. <u>Compliance Guidelines for Specific Provisions of 29 CFR 1910.119</u>. Guidelines for assessing compliance with the provisions of the PSM standard are provided in Appendix A of this instruction.

    1. CSHOs shall use the guidance contained in Appendix A during all enforcement activities related to the PSM standard.

    2. Clarifications and interpretations are provided in Appendix B of this instruction. Appendix B (or a subsequent revision) shall normally be the first point of reference in interpreting 29 CFR 1910.119.

        NOTE: Appendix B will be updated on an ongoing basis through page changes to this instruction, as more interpretations are developed. CSHOs must therefore take care to ensure that their reference copies are up-to-date.

N. <u>Citations</u>. Citations for violations of the PSM standard shall be issued in accordance with the FOM, Chapters IV and V, with the following additional directions:

    1. <u>Classification</u>. The requirements of the PSM standard are intended to eliminate or mitigate the consequences of releases of highly hazardous chemicals. The provisions of the standard present closely inter-related requirements, emphasizing the application of management controls when addressing the risks associated with handling or working near hazardous chemicals.

      a. Any violation of the PSM standard, therefore, is a condition which could result in death or serious physical harm to employees.

      b. Accordingly, violations of the PSM standard shall normally not be classified as "other-than-serious."

348 OSHA and EPA PSM Requirements

OSHA Instruction CPL 2-2.45A

Directorate of Compliance Programs

2. **Use of Appendix A.** Appendix A, PSM Audit Guidelines, is constructed as a series of questions relating to each of the pertinent provisions of the standard.

   a. The questions are designed to elicit a determination of "Yes" or "No" by the CSHO as to whether compliance with the provision has been met.

   b. A determination of "No" for any provision indicates noncompliance; thus, any "No" shall normally result in a citation for a violation of that provision.

   c. The CSHO shall thoroughly document each such determination in the case file.

O. **Non-Mandatory Appendices to this Instruction.** This instruction contains two **non-mandatory** appendices that are designed to provide additional compliance assistance.

   1. Appendix E, Recommended Health Care Management Program Components for Process Safety Management, is still being developed and is designated as "Reserved."

   2. Appendix G, Recommended Guidelines for PQV Inspection Preparation, is intended as an aid to Regional and Area Offices in planning resources for PQV and other PSM-related inspections.

P. **Evaluation.** Each Region shall develop a preliminary evaluation of the effectiveness of this program and submit it to the Directorate of Compliance Programs no later than September 30, 1993. The report shall include, at a minimum, the following items:

   1. The utility of the PSM audit guidelines used in Appendix A.

   2. An assessment of the accuracy of targeting information.

   3. An estimate of total resources (CSHO, supervisory, administrative and legal) that were required to conduct each inspection.

Appendix 9 - OSHA Compliance Directive 349

OSHA Instruction CPL 2-2.45A

Directorate of Compliance Programs

Q. **Recording in IMIS**. Information about PSM-related inspections shall be recorded in IMIS following current instructions given in the IMIS manual. Refer to Appendix H of this instruction for additional guidance.

[signature]

Dorothy L. Strunk
Acting Assistant Secretary

DISTRIBUTION: National, Regional, and Area Offices
All Compliance Officers
State Designees
NIOSH Regional Program Directors
7(c)(1) Consultation Project Managers
OSHA Training Institute

350 OSHA and EPA PSM Requirements

OSHA Instruction CPL 2-2.45A

Directorate of Compliance Programs

APPENDIX A

PSM AUDIT GUIDELINES

Purpose.

This appendix contains audit guidelines intended to assist the CSHO in investigating an employer's compliance with the PSM standard. It shall be used in conjunction with Appendix B, Clarifications and Interpretations of the PSM Standard, as the primary source of compliance guidance on 29 CFR 1910.119.

Structure.

The guidelines present a Program Summary, Quality Criteria References, and a Verification checklist for each of the PSM elements.

1. Guidelines for paragraphs c, g, h, k, m, and n are designed so that CSHOs who may not be specifically trained in chemical process plants or in the PSM standard can make a preliminary review of the required elements.

2. Guidelines for elements d, e, f, i, j, l, o, and p are oriented toward more detailed investigations.

Use of the Verification Checklist.

The verification of each program element is divided into three parts: *Records Review, On-Site Conditions* and *Interviews*.

1. The *Records Review* section describes the documentation of the programs as required by the PSM standard. During a preliminary inspection, the CSHO shall review the documentation for the entire PSM program to ascertain that all of the elements are developed.

2. Sections labeled *On-Site Conditions* and *Interviews* guide the CSHO in confirming that the programs are implemented. This confirmation involves *observing* conditions and procedures, and *interviewing* the operators, maintenance personnel, engineering support staff, contractors and contractor employees, as appropriate, to determine whether the implemented program matches the program outlined by the documentation.

Appendix 9 - OSHA Compliance Directive

OSHA Instruction CPL 2-2.45A

Directorate of Compliance Programs

> NOTE: Several questions in the "Interviews" sections refer to interviewing *engineers*. The PSM standard does not require an employer to employ engineers, and these questions should not be construed as imposing a new requirement that an employer do so. All questions in this appendix that refer to interviews of engineers shall be understood to mean "engineers, if any, or other qualified persons capable of providing the information requested."

3. The CSHO shall initially perform a representative number of observations and interviews for elements c, g, h, k, m, and n. A more detailed investigation will cover all 14 elements. During these detailed assessments, the CSHO shall review components from a representative number of processes, if multiple processes exist. To confirm implementation, the CSHO shall compare the conditions and the interview results with both the minimum requirements of the PSM standard and the program outlined by the employer's documents.

Audit Guideline Documentation.

As noted at P.2. of the body of this instruction, the Audit Guidelines are constructed as a series of questions relating to each of the pertinent provisions of the standard.

1. The questions are designed to elicit a determination of "Yes" or "No" by the CSHO as to whether compliance with the provision has been met. This shall be indicated in the column labeled Met Y/N. A "Y" or "Yes" in this column indicates the subsection meets requirements. An "N" or "No" indicates the employer does not meet the standard and an "NA" signifies that the subsection does not apply.

2. A determination of "No" for any provision indicates noncompliance; thus, **any "No" shall normally result in a citation for a violation of that provision.**

3. The CSHO shall thoroughly document each such determination in the case file.

The *Field Note Reference(s)* space is used to cross-reference the PSM subsection with the CSHO's field notes. Field notes need not be rewritten when using these guidelines. The CSHO may record field note page numbers, videotape frame identification, photograph identification, and other documentation that refers to the requirements of the standard's elements.

352 OSHA and EPA PSM Requirements

OSHA Instruction CPL 2-2.45A

Directorate of Compliance Programs

Basic Audit Information.

In order to gather the information needed to audit the program, the CSHO shall answer the following questions for each element:

Who? What? When? Where? Why? and *How?*

1. *Who* are the officials responsible for developing and implementing each of the program elements?
2. *What* are the requirements and the contents of each program element?
3. *When* are the required actions for each element completed and when are they required to be completed?
4. *Where* have actions been implemented or changed?
5. *Why* have the implementation decisions and priorities been made as recorded in the PSM documentation?
6. *How* is the program implemented and how is the program's effectiveness evaluated and improved (monitoring performance, followup and closure of outstanding items, etc.)?

Interrelationship of Elements.

An essential part of verifying program implementation is to audit the flow of information and activities among the elements. When information in one element is changed or when action takes place in one element that affects other elements, the CSHO shall review a sample of the related elements to see if the appropriate changes and followup actions have taken place.

The following example demonstrates the interrelationship among the elements:

During a routine inspection of equipment (**Mechanical Integrity**), the maintenance worker discovers a valve that no longer meets the applicable code and must be changed. Because the type of valve is no longer made, a different type of valve must be selected and installed (**Management of Change**). The type of valve selected may mandate different steps for the operators (**Operating Procedures**) who will require training and verification in the new procedures (**Training**). The rationale for selecting the type of valve

Appendix 9 - OSHA Compliance Directive 353

OSHA Instruction CPL 2-2.45A

Directorate of Compliance Programs

must be made available for review by employees and their representatives (**Employee Participation**).

When the new valve is installed by the supplier (**Contractors**), it will involve shutting down part of the process (**Pre-startup Safety Review**) as well as brazing some of the lines (**Hot Work Permit**). The employer must review the response plan (**Emergency Planning**) to ensure that procedures are adequate for the installation hazards.

Although **Management of Change** provisions cover interim changes, after the new valve is in place the **Process Safety Information** will have to be updated before the **Process Hazard Analysis** is updated or revalidated, to account for potential hazards associated with the new equipment. Also, inspection and maintenance procedures and training will need to be updated (**Mechanical Integrity**).

In summary, 11 PSM elements can be affected by changing one valve. A CSHO would check a representative number of these 11 elements to confirm that the required followup activities have been implemented for the new valve.

Three key elements shall be routinely reviewed to verify that changes have been implemented. They are:

- Operating Procedures;
- Process Hazard Analysis; and
- Training.

These elements shall be crosschecked to see if they show that the changes have been followed through to completion.

354 OSHA and EPA PSM Requirements

OSHA Instruction CPL 2-2.45A

Directorate of Compliance Programs

Appendix B

Clarifications and Interpretations of the PSM Standard

The guidance contained in this appendix is provided for compliance assistance. It shall be followed in interpreting the PSM standard for compliance purposes. Unless otherwise noted, all paragraph citations refer to 29 CFR 1910.119.

Where possible, clarifications and interpretations have been presented in a question-and-answer format.

NOTE: OSHA plans to include additional clarifications and interpretations in this appendix through future page changes to this instruction.

(a) Application

(a) Registration

Do covered establishments have to register with OSHA?

No. There is no requirement that establishments covered by the standard register with or otherwise notify OSHA.

(a) Explosives--fireworks manufacture

How does the PSM standard apply to pyrotechnics (fireworks) and explosives?

The PSM standard amended the scope of 29 CFR 1910.109, Explosives and blasting agents, by revising paragraph (k), which requires that the manufacturer of explosives and pyrotechnics comply with 29 CFR 1910.119. As defined at 1910.109(a)(10), pyrotechnics are commonly referred to as fireworks. Employers who manufacture explosives and fireworks must comply with both 29 CFR 1910.109 and 1910.119.

The applicability of 29 CFR 1910.109 to employers who manufacture fireworks is delineated in OSHA Instruction

Appendix 9 - OSHA Compliance Directive 355

OSHA Instruction CPL 2-2.45A

Directorate of Compliance Programs

CPL 2.73, Fireworks Manufacturers: Compliance Policy. In accordance with that directive, a fireworks plant employer can be cited for violation of 29 CFR 1910.109 with reference to certain National Fire Protection Association (NFPA) standards in NFPA 1124, Code for the Manufacture, Transportation and Storage of Fireworks.

What is the role of the Bureau of Alcohol, Tobacco and Firearms (BATF) vis-a-vis the PSM standard and fireworks manufacture?

By 27 CFR 55 Subpart K, BATF regulates the storage, including minimum distances, of explosive materials including fireworks in the workplace. As such, BATF limits the amount of special fireworks, pyrotechnic composition, and explosive materials used to assemble fireworks in processing building to no more than 500 pounds. Also, the maximum quantity of flash powder permitted by BATF in any fireworks process building is 10 pounds. These BATF limitations should not be confused with the applicability of the PSM standard to _any_ amount of fireworks being manufactured.

(a) Laboratories

Does the PSM standard apply to laboratory and research operations?

A laboratory or research operation involving at least the threshold quantity of one or more highly hazardous chemicals is subject to the PSM standard.

(a) Flammable liquids

Are processes involving flammable liquids (e.g., ethyl alcohol) covered by the standard?

Processes involving flammable liquids (e.g., in a distillation process) in quantities at or above 10,000 lbs. are covered. Quantities of flammable liquids in storage are considered a part of the process if the storage tanks are interconnected with the process, or if they are sufficiently near the process that an explosion, fire, or release could reasonably involve the storage area combined with the process in quantities sufficient to meet the threshold amount of 10,000 lbs.

356 OSHA and EPA PSM Requirements

OSHA Instruction CPL 2-2.45A

Directorate of Compliance Programs

Flammable liquids that are stored on a tank farm (e.g., a wholesale gasoline regional tank farm) where only transferring and storage are done are not covered by the PSM standard. They are, however, covered under 1910.106.

(a)(1)(ii)(A)   Tote tanks

350-gallon tote tanks containing flammable liquids are used at a facility to refuel vehicles. Are they covered by the standard?

No. 1910.119(a)(1)(ii)(a) exempts hydrocarbon fuels used solely for workplace consumption as a fuel (e.g., gasoline for vehicle refueling) if such fuels are not part of a process containing another highly hazardous chemical covered by the standard. They are, however, covered under 1910.106.

(a)(2)(i)   Retail facilities

What is the definition of "retail facilities" that are exempted from coverage by the PSM standard?

With respect to enforcement of the PSM standard, a _retail facility_ means an establishment that would otherwise be subject to the PSM standard at which more than half of the income is obtained from direct sales to end users.

(a)(2)(iii)   Meaning of "facility"

Can a facility contain more than one process?

A facility can include multiple processes. If multiple processes are interconnected, they may be considered a single process for purposes of the standard.

Appendix 9 - OSHA Compliance Directive

OSHA Instruction CPL 2-2.45A

Directorate of Compliance Programs

(b) Definitions

(b) "Process"

What are "aggregate threshold quantities"?

> In accordance with the second sentence of the definition of "process," quantities of a particular hazardous chemical contained in vessels that are interconnected--and in unconnected vessels that may be adversely affected due to an incident at a nearby process--may be combined to determine whether the threshold level of a hazardous chemical has been reached. If the threshold level is exceeded by the combination of the amount in separate tanks and interconnected vessels, then all of these may be considered one process.

(b) "Hot work"

> **"Spark-producing operations"** include operations which use flame- or spark-producing equipment--such as grinders, welding, burning, or brazing--that are capable of igniting flammable vapors or gases.

(c) Employee participation

(c)(2) Consultation

What does consult mean? Can the employer simply inform the employees?

> The intent of consult is to exchange information, solicit input and participation from the employees and their representatives. It requires more than simply informing employees.
>
> The employer needs to consult with employees and employee representatives and develop information concerning knowledge and expertise of individual employees in various processes and aspects of the facility in order to ensure substantive input by employees and their representatives in developing

358 OSHA and EPA PSM Requirements

OSHA Instruction CPL 2-2.45A

Directorate of Compliance Programs

the written action plan, process hazard analyses, and access to information required under the standard.

The standard requires employers to consult with "employees and their representatives." Is the term broad enough to include a representative of the international union? A consultant designated by the union local or international?

The standard requires consultation with "employees and their representatives". The term "employee representative" is intended to mean union representative where a union exists, or an employee-designated representative in the absence of a union. The term is to be construed broadly, and may include the local union, the international union, or an individual designated by these parties, such as the safety and health committee representative at the site or a non-employee consultant. In the absence of a union, employees have a right under the standard to designate a representative to participate in the consultation process.

(c)(3)   Access

What does "access" mean? Does this mean simply make it available at a central location? Does the employer have to make copies for employees if requested?

The intent of access under this standard is for the information to be made available for employees and their representatives in a reasonable manner. Reasonable access may require providing copies or loaning documents. The trade secret provision of the standard permits the employer to require confidentiality agreements before providing the information.

(d)   Process safety information

(d)   Retention of information

How long must the employer maintain process safety information?

In order to demonstrate compliance with this paragraph, and to meet the purpose of the standard, the process safety information is to be kept for the lifetime of the process, and updated whenever changes other than "replacement in kind" are made.

Appendix 9 - OSHA Compliance Directive

OSHA Instruction CPL 2-2.45A

Directorate of Compliance Programs

(e) Process hazard analysis

(e)(1) PHA priority

What rationale must employers use to determine the priority for conducting the process hazard analyses? May the rationale include age, history, extent of employee exposure, etc.?

> The appropriate priority for conducting PHA's is to be determined by using all of the criteria identified in this paragraph, e , extent of the process hazards (catastrophic potential), age of the process, number of potentially exposed employees, and operating history. Other appropriate factors may also be considered in establishing the priority. The documentation required by this paragraph shall demonstrate the underlying rationale for the prioritization.

(e)(5) Timeliness

Employers must "promptly" address the problems identified in the PHA in a "timely manner," and complete actions "as soon as possible." What time frame did OSHA intend here?

> The standard's intent is for the employer to take corrective action as soon as possible. _As soon as possible_ means that the employer shall proceed with all due speed, considering the complexity of the recommendation and the difficulty of implementation. OSHA expects employers to develop a schedule for completion of corrective actions and to document the basis for the amount of time needed.

(e)(7) Retention

How long must the process hazard analyses, updates, and revalidations be retained?

> For the life of the process.

360 OSHA and EPA PSM Requirements

OSHA Instruction CPL 2-2.45A

Directorate of Compliance Programs

| (f) Operating procedures |

(f)(1)(iii)(c) "Control measures to be taken if physical contact or airborne exposure occurs"

Does this mean **first aid**, or **industrial hygiene** services?

It primarily means first aid procedures or emergency medical attention, which should be consistent with the information on the material safety data sheet.

| (g) Training |

(g)(3) Training documentation

This paragraph requires the employer to make sure that operators "understand" the training provided to them under this section. Is some method of testing required?

There must be some positive means taken by the employer to determine if employees have understood their training and are capable of adhering to the current operating procedures of the process. This could include the administration of a written test, although the standard does not require that a formal written test be used. Other means of ascertaining comprehension of the training, such as on-the-job demonstrations, etc., are acceptable, as long as they are adequately documented.

| (j) Mechanical integrity |

(j)(2) Written procedures

Do these written procedures need to be specific to each vessel, each type of vessel, or each group of equipment types listed?

The procedures need to be specific to the type of vessel or equipment. Identical or very similar vessels and items of equipment in similar service need not have individualized maintenance procedures. Each procedure must clearly identify the equipment to which it applies.

Appendix 9 - OSHA Compliance Directive

OSHA Instruction CPL 2-2.45A

Directorate of Compliance Programs

(j)(6)(ii) Quality assurance

If an installation is being done by contractors, does this require the employer to implement a quality assurance program to monitor the activities of these contractors?

> The employer is responsible for ensuring that equipment is installed consistent with design specifications and manufacturer's instructions. This may require the employer to be involved in the review, inspection, certification, and quality assurance of work performed by contractors.

362 OSHA and EPA PSM Requirements

OSHA Instruction CPL 2-2.45A

Directorate of Compliance Programs

Appendix C

**Standard Industrial Classification (SIC) Codes
Targeted for PQV Inspections**

The following SIC codes are designated as targeted for possible PQV inspections, in accordance with the criteria at J.1. of this instruction:

- 2812 Alkalies and Chlorine
- 2819 Industrial Inorganic Chemicals, Not Elsewhere Classified
- 2821 Plastics Materials, Synthetic Resins, and Nonvulcanizable Elastomers
- 2865 Cyclic Organic Crudes and Intermediates, and Organic Dyes and Pigments
- 2869 Industrial Organic Chemicals, Not Elsewhere Classified
- 2873 Nitrogenous Fertilizers
- 2892 Explosives [chemical plants making explosives]
- 2911 Petroleum Refining

Appendix 9 - OSHA Compliance Directive

OSHA Instruction CPL 2-2.45A

Directorate of Compliance Programs

Appendix D
References for Compliance with the PSM Standard

1. OSHA Instruction CPL 2.45B, June 15, 1989, the Revised Field Operations Manual (FOM).
2. OSHA Instruction STP 2.22A, CH-2, January 29, 1990, State Plan Policies and Procedures Manual.
3. OSHA Instruction ADM 1-1.12B, Dec. 29, 1989, Integrated Management Information System (IMIS) Forms Manual, Chapter V.
4. OSHA Instruction CPL 2-2.45, Sep. 6, 1988, Systems Safety Evaluation of Operations with Catastrophic Potential.
5. "Safety and Health Program Management Guidelines," 1989; U.S. Department of Labor, Occupational Safety and Health Administration.
6. "Safety and Health Guide for the Chemical Industry," 1986, (OSHA 3091); US.DOL, OSHA.
7. "Review of Emergency Systems," June 1988; U.S.E.P.A., Office of Solid Waste and Emergency Response, Washington, DC 20480.
8. "Guidelines for Hazard Evaluation Procedures," Center for Chemical Process Safety of the American Institute of Chemical Engineers; 345 East 47th Street, New York, NY 10017.
9. "Plant Guidelines for Technical Management of Chemical Process Safety," Center for Chemical Process Safety (CCPS) of The American Institute of Chemical Engineers (AICHE).
10. "Guidelines for Safe Storage and Handling of High Toxic Hazard Materials," AICHE, CCPS.
11. "Guidelines for Vapor Release Mitigation," AICHE, CCPS.
12. "Process Safety Management (Control of Acute Hazards)," Chemical Manufacturers Association (CMA).
13. "Evaluating Process Safety in the Chemical Industry," Chemical Manufacturers Association; 2501 M Street NW, Washington, DC 20037.

364 OSHA and EPA PSM Requirements

OSHA Instruction CPL 2-2.45A

Directorate of Compliance Programs

14. "Safe Warehousing of Chemicals," Chemical Manufacturers Association.

15. "A Managers Guide to Reducing Human Errors Improving Human Performance in the Chemical Industry," Chemical Manufacturers Association.

16. "Improving Owner and Contractor Safety Performance," API Recommended Practice 2220.

17. "Management of Process Hazards," American Petroleum Institute (API) Recommended Practice 750, First Edition, January 1990; 1220 L Street NW, Washington, DC 20005.

18. "Sizing, Selection, and Installation of Pressure Relieving Devices," Part 1, July 1990, API RP 520.

19. "Guide for Pressure relieving and Depressuring Systems," Nov. 1990, API RP 521.

20. "Avoiding Environmental Cracking in Amine Units," Aug. 1990, API RP 945.

21. "Pressure Vessel Inspection Code: Inspection, Rating, Repair, and Alteration," June 1989, API STD 510.

22. "Inspection of Piping, Tubing, Valves, and Fittings," API RP 574.

23. "Prevention of Brittle Fracture of Pressure Vessels," API RP 920.

24. "Accident Investigation * * * A New Approach," 1983, National Safety Council; 444 North Michigan Avenue, Chicago, IL 60611-3991.

25. "Fire & Explosion Index Hazard Classification Guide," 6th Edition, May 1987, Dow Chemical Company; Midland, Michigan 48674.

26. "Chemical Exposure Index," May 1988, Dow Chemical Co.

27. "Pressure Vessels, Section VIII," The American Society of Mechanical Engineers (ASME).

28. "Chemical Plant and Petroleum Refinery Piping," ASME B31.3.

Appendix 9 - OSHA Compliance Directive 365

OSHA Instruction CPL 2-2.45A

Directorate of Compliance Programs

29. "Personnel Qualification and Certification in Nondestructive Testing," American Society of Nondestructive Testing, Recommended Practice No. SNT-TC-1A.

30. "Prevention of Furnace Explosions/Implosions in Multiple Burner Boiler Furnaces," National Fire Protection Association, NFPA 85C.

31. "Purged and Pressurized Enclosures for Electrical Equipment," NFPA 496.

32. "Spacing of Facilities in Outdoor Chemical Plants," Factory Mutual Loss Prevention Data Sheet, 7-44.

33. "Chemical Process Control and Control Rooms," Factory Mutual Loss Prevention Data Sheet, 7-45.

34. "National Board Inspection Code, A Manual for Boiler and Pressure Vessel Inspectors," The National Board of Boiler and Pressure Vessel Inspectors, 1992.

35. Gideon, James A., and Thomas W. Carmody, "Process Safety Management: Resources from the American Institute of Chemical Engineers for Use by Industrial Hygienists," American Industrial Hygiene Association Journal (53), June 1992.

366 OSHA and EPA PSM Requirements

OSHA Instruction CPL 2-2.45A
Directorate of Compliance Programs

Appendix E

Recommended Health Care Management Program Components
For Process Safety Management
(Nonmandatory Guidance)

[RESERVED]

Appendix 9 - OSHA Compliance Directive 367

OSHA Instruction CPL 2-2.45A

Directorate of Compliance Programs

Appendix F

SAMPLE LETTER TO BE PROVIDED TO EMPLOYER FOLLOWING SCREENING FOR PSM COVERAGE

Dear Employer:

The Occupational Safety and Health Administration's (OSHA) evaluation of your workplace and the information you have provided indicate that your establishment is covered by OSHA's standard for Process Safety Management of Highly Hazardous Chemicals, 29 CFR 1910.119.

As you may know, the requirements of this standard are intended to protect employees by preventing or minimizing the consequences of accidents involving highly hazardous chemicals. OSHA believes that compliance with the standard is important for ensuring worker protection.

Accompanying this letter are copies of the following publications:

(1) "Process Safety Management," OSHA 3132, which provides information on the standard and its requirements, and

(2) "Process Safety Management--Guidelines for Compliance," OSHA 3133.

These guidelines are not a substitute for the standard itself, which is reprinted in the "Process Safety Management" booklet.

Your workplace will be included in an overall listing of all establishments known to be covered by the standard. OSHA may schedule an inspection of your establishment in the future to evaluate your compliance with the standard.

This standard is relatively new, and OSHA realizes that affected employers may have many questions. After reading the materials provided, please feel free to direct any questions to [Area Office; telephone number].

Sincerely,

Area Director

368 OSHA and EPA PSM Requirements

OSHA Instruction CPL 2-2.45A

Directorate of Compliance Programs

Appendix G

Recommended Guidelines for PQV Inspection Preparation
(Nonmandatory)

The following guidelines are suggested as background and preparation for a PQV inspection.

AREA OFFICE COORDINATION

Coordination within the Area Office is absolutely essential in the orderly conduct of a PQV inspection. The Area Director and all those involved in a PQV inspection must commit the resources with the understanding that the project is long-term, possibly several weeks or months. It is imperative that team members complete all outstanding assignments prior to the PQV inspection. Equally important, team participants should not be directed or "asked" to do assignments while they are engaged in the PQV inspection. An obvious exception would be court hearings, over which the Area Office has little control.

The Area Director should designate a contact person in the Area Office to coordinate and oversee all aspects of the inspection. The contact person should be a supervisor, safety or industrial hygiene (IH), who is familiar with the PQV concept. In addition to providing Area Office coordination, the contact person would review the entire case file/report. The team leader would communicate at least weekly with the contact supervisor, who would then brief the Area Director as appropriate. The contact supervisor would advise the supervisor(s) of team members about the conduct and performance of the individuals concerned. This is especially important so that the team members get a fair appraisal from their supervisors who would not or could not rate them on their PQV performance because they had no knowledge as to what was going on for weeks or months.

INSPECTION TEAM COMPOSITION

By design, a PQV inspection is a large and complex undertaking, to be accomplished by a select, well-trained team. All members of the team must be experienced journey or senior level compliance officers who are familiar with the chemical industry and have taken the appropriate OSHA training. Newer compliance officers can be utilized in the inspections, but not as a substitute for regular team members.

The team should consist of two safety compliance officers/engineers, two industrial hygiene compliance officers, an administrative support person and a construction specialist. The team leader could be from either discipline in the team, but preferably a safety specialist, due to the fact that most of the critical PSM and construction related hazards reside in the area of safety.

The team leader should be a GS-12 Senior Compliance Officer with experience in large team inspections. He or she should have excellent organizational and communication skills, both oral and written. It would also be of benefit that the team leader be knowledgeable in word processing and data base management computer operations. Since the team leader will be the focal point during the conduct of the inspection, that person should also have demonstrated leadership abilities. The entire team, the company, employees/unions and other OSHA personnel will look to the team leader for direction and answers to the many questions that will arise during the course of the inspection.

Appendix 9 - OSHA Compliance Directive

OSHA Instruction CPL 2-2.45A

Directorate of Compliance Programs

The team leader is responsible for the overall conduct of the inspection including planning, onsite activities and report preparation. The leader would assign the various inspection areas to team members in accordance with their expertise and abilities, and determine what, if any, special expertise is needed. Additional responsibilities include:

1. Keeping the Area Office contact apprised of activities;
2. Providing and tracking requests for documents;
3. Resolving problems with the company;
4. Ensuring that the report addresses all questions in the directive.

An administrative support person would greatly increase the overall efficiency of the inspection. This position would be ideal for an accommodated compliance officer with some computer skills and organizational abilities. The support person would answer directly to the team leader and would be responsible for organizing, labeling and filing the many documents that will become part of the case file. An accommodated CSHO could also review the documents and document requests to assure the request was properly fulfilled. In addition, an accommodated CSHO could assist the inspecting team members with the many interviews that will be conducted. The support person would also be responsible for the inspection supplies and equipment.

Safety and IH team members are responsible for carrying out the PQV inspection activities under the direction of the team leader. They must keep the team leader apprised of their activities and potential problems when they arise. The construction specialist would work for the most part independently of the rest of the team, under the general direction of the team leader. Some crossover of inspection areas is to be expected, as many of the contractors and company responsibilities overlap.

PRE-INSPECTION PREPARATION

Effective planning and preparation is essential to the efficient implementation and successful completion of any large inspection, especially a PQV. Exhibit 1 provides an outline that can be used as a guide to plan and prepare for a PQV inspection. Establishment histories can be obtained and reviewed well in advance of the target date for the inspection. The inspection strategy and scheduling should be done after the team has been selected. A pre-inspection meeting with all members and the Area Office contact person should be held prior to entry.

The case file begins in the planning and preparation stage. Any documents received, such as micro to host reports, citations and PSM-related findings (including PetroSEP) in other regions must be logged and identified to allow for easy retrieval. An activity log/diary should be started to record all pertinent actions taken. A computer data base management program is recommended to keep track of the document requests and to provide a ready index of the documents that have been obtained. With this type of system it would be easy to search for pertinent documents by using the OSHA identification number, topic of document, company identification number, date of request, etc., and to ensure that various members of the team do not duplicate requests for documents.

The team should develop a weekly schedule of activities, taking into account travel days, holidays, start time, stop time, company briefings and internal briefings. Time should be allotted during the inspection week to complete necessary paperwork and documentation and tie up loose ends.

370 OSHA and EPA PSM Requirements

OSHA Instruction CPL 2-2.45A

Directorate of Compliance Programs

DOCUMENTS

PQV inspections will require compliance officers to review numerous company documents. Many of these documents will become part of the case file as documentation for potential citations or for documentation of the required PSM elements. It is imperative that these documents be organized and identified so that they may be readily referenced and reviewed. It is highly recommended that all requests for copies of company documents be in writing. A standard document request format should be established and should contain at least the following information:

1. Who is the requester
2. To whom the request is made
3. Identity of the document (in company terms if possible)
4. Company document number
5. Date of request
6. Priority for response
7. Internal I.D. number or docket number (for filing)
8. Date request fulfilled
9. Comment section (did the response fulfill request).

It should be noted that there is no universal language used to identify documents. Different companies have different names for the same type of document. It is therefore essential to clearly communicate what information is needed and desired prior to writing the request. The document requests should be in duplicate: one copy for the company and the other to be retained in the case file. To avoid long discussions and legal department involvement, all documents obtained should be considered proprietary information.

Prior to the documents actually being received, a filing system should be developed. The system should be secure, accessible to all team members and ensure that individual documents are easily retrievable. The administrative support person could manage the filing system to ensure its continued effectiveness. NOTE: Only appropriate documents should be maintained in the filing system; field notes, document "clips", and document review/evaluation notes should remain with the corresponding 1B's.

Exhibit 2 contains a list of those documents most commonly requested. It is divided into two sections: Pre-Unit Selection and Unit-Specific Documents.

INSPECTION FACILITIES

The PQV team needs a suitable work area/command center from which the inspection can be conducted and coordinated. Except in the most unusual of circumstances, the company will provide the requisite onsite space. Almost any room will suffice, providing it meets some basic requirements. The work area must be secure 24 hours a day with access limited to the inspection team and those company officials who would respond in an emergency. This is important so as to preclude taking boxes of documents and equipment in and out each day. The room should have sufficient desks and/or tables for reviewing documents and writing the report. Provisions should be made for communications--one phone line as a minimum. Outgoing calls should be charged on the Area Office calling card. Where phone service is not provided, the team should have a portable cellular phone.

Appendix 9 - OSHA Compliance Directive

OSHA Instruction CPL 2-2.45A

Directorate of Compliance Programs

Sufficient power outlets should be available for charging pumps, batteries and other inspection equipment.

The inspection team will need copies of a number of documents. It is hoped that the company would provide copying services or the use of a copy machine. The administrative support person could make the copies should the company not provide these services.

The team leader must determine as soon as possible, what--if any--of the necessary facilities the company will provide. If the company does not provide all of the necessary facilities voluntarily, or puts disruptive restrictions on their use, the ARA for Technical Support should be contacted as soon as possible so that alternate facilities can be arranged. This may result in the use of a rented copier(s) and office space.

INSPECTION EQUIPMENT

Upon entry to the site, the inspection team should be fully prepared with all necessary inspection equipment and personal protective equipment. Exhibit 3 contains a list of equipment that may be useful to prepare for the inspection. In addition, an inspection "kit" is outlined which can be used to set up the command center. Some of the items in the "kit" may appear to be trivial; however, all of these items will be needed at some time during the inspection. It may not be practical to go back and forth to the office or a store to get these items, particularly if the inspection site is in a remote location. The administrative support person would be responsible for maintaining adequate supplies throughout the inspection.

CRITICAL INSPECTION AREAS

It is essential that team members have specific subjects and areas to investigate. The team leader, with input from the team members, should assign the inspection areas prior to entry. This will help to avoid confusion and duplication of effort. In addition, the team members will be able to be better prepared for their individual tasks.

As inspection subjects are completed, the information should be reviewed with the team leader before going on to the next assignment. The state of compliance or noncompliance within any given area may require the team leader to modify the assignment list so as to make the most of the resources available.

CONTRACTORS

Contractors are an integral part of any PSM inspection. There may be only a few contractors or dozens, with several hundred contract employees, depending on whether the facility is undergoing a shutdown or turnaround.

It is imperative that, upon entry, the scope of the contractor activity be determined. The construction specialist on the team will have to formulate an inspection plan and set appropriate priorities. It is not the intent of the PQV inspection to inspect all outside contractors that are on-site, rather to inspect only those contractors who may be exposed to, or could cause or be affected by a catastrophic incident. Food service workers, certain janitorial employees and similar activities would

372 OSHA and EPA PSM Requirements

OSHA Instruction CPL 2-2.45A

Directorate of Compliance Programs

not normally be inspected. Remote construction projects not associated with catastrophic potential would not necessarily be inspected.

The term "contractor" is not limited to construction type activities. Many chemical facilities use contract maintenance workers, vessel and piping inspectors, vessel heat treating, cleaning, engineering and similar non-construction contractors who remain at the facility year round or are called in at regular intervals. They are used to supplement existing plant personnel for regular duties and for special projects.

A shared responsibility for both contractors and company is quality assurance. It is essential that all materials and workmanship meet engineering standards. There should be sufficient checks to ensure that materials, such as the proper alloy or carbon steel pipe is used, and that the studs and/or bolts are of the proper size and grade. This is especially important in contractor supplied materials.

CRITICAL EXPERTISE

Situations may arise in a PQV inspection that are beyond the technical expertise of the team members. A list should be developed identifying OSHA personnel and/or private sector experts and how they may be contacted. Areas where this expertise may be needed are:

1. Pressure equipment
2. Fire protection (fire brigades)
3. Facility siting
4. Emergency medical services
5. Hazardous waste operations
6. Dispersion modeling & incident command centers
7. Process hazard analysis/HAZOPS
8. Process chemistry
9. Industry practice

By no means is this list all-inclusive. It should be modified as needed to reflect current technology and hazards.

DOCUMENTATION

In order to withstand the probable legal challenges, all items must be thoroughly documented. Since the team will be made up of journey- and senior-level CSHOs, good documentation is to be expected. All OSHA-1B forms must be complete and legible. Shortcuts for employer knowledge such as "should have known" or "reasonable diligence" are not acceptable. Appropriate company documents, logs, procedures, permits, etc., should be referenced on the 1B for the particular violation.

Photographic documentation, either still camera or videotape, should be reviewed as soon as practicable to ensure that the condition or violation is appropriately depicted. Retake any photos or videos that are not good quality.

CASE FILE AND REPORT PREPARATION

Appendix 9 - OSHA Compliance Directive

OSHA Instruction CPL 2-2.45A

Directorate of Compliance Programs

A PQV inspection will take weeks or months of onsite activity and will generate a large amount of paper, both in field notes and documents. It is essential that the paper flow be organized and well maintained. This will result not only in a more efficient onsite survey, but will greatly reduce the write-up time.

A daily log, either manual or computer generated, should be maintained indicating the team members onsite, daily activities, meetings, problems, or other details, as necessary. All OSHA-1B forms should be completed as the violation is observed, documenting the employees exposed, the date, time, location and management representative who accompanied the CSHO. Each instance of a violation should have a separate 1B. Where multiple violations are noted on a form, the form should be photocopied and highlighted showing the appropriate instance and corresponding documentation. Alleged violation descriptions should be written as soon as practicable, while the hazard is fresh in the mind of the CSHO. Multi-employer policy citations must be coordinated with respect to exposing, controlling, correcting and creating employers' files.

Case file structure and organization must begin prior to entry into the facility. All documents must be logged and an index (computer preferred) generated, indicating the subject matter, document identification number, file number and the location of the document (box number). This is essential, as these documents may have to be referenced or retrieved many times during the course of the inspection and the review process. A data base management program for the PC's would be extremely beneficial. Computer disks should be backed up daily, or more often as necessary. The photos and videotape taken during the inspection should be properly identified with photographer, date, roll or tape number and subject. They should be kept in a separate file.

374 OSHA and EPA PSM Requirements

OSHA Instruction CPL 2-2.45A

Directorate of Compliance Programs

EXHIBIT 1

PRE-INSPECTION PREPARATION

A. Previous OSHA history - nationwide search

    1. all citations and/or reports
    2. litigation results
    3. outstanding issues, items in contest
    4. health response team reports
    5. NIOSH evaluations if any

B. EPA history
    1. reportable releases
    2. reports of any kind
    3. complaints and pending actions

C. Other Agency histories - local/State/Federal

    1. Dept. of Transportation
    2. Coast Guard
    3. ESDA/FEMA
    4. State Fire Marshal
    5. State Boiler and Pressure Vessel

D. Previous PetroSEP/PSM inspection results

    1. citations
    2. team members & expertise
    3. settlement agreements or litigation results

E. Identify contact people -- other jurisdictions

- EPA, DOT, Coast Guard, etc.

F. Acquire necessary codes or standards

- ASME, API, ANSI, NFPA, etc.

INSPECTION STRATEGY

A. Identify critical needs and expertise

B. Select team members

C. Identify expertise within the team

Appendix 9 - OSHA Compliance Directive

OSHA Instruction CPL 2-2.45A

Directorate of Compliance Programs

D. Identify critical inspection areas

E. Assign areas according to expertise

F. Identify areas lacking expertise

    1. provide training
    2. bring in additional resources

G. Develop a tracking system for documents

H. Develop a daily log of on-site activities

I. Identify known scheduling conflicts

- Team members and/or employer

J. Develop weekly schedule of activities

    1. travel, write up, start/stop times
    2. employer/employee and Area Office updates

PLANNING AND SCHEDULING

A. Create a Projected Time Line

    1. Projected records and program review time
    2. Projected walkaround time
    3. Projected write-up time

B. Resource Scheduling

    1. Team leader and construction specialist enter first for program and records review; present document request list.
    2. Full team enters following acquisition of requested documents for program/record review & walkaround
    3. Expert assistance enters as needed

C. Equipment Acquisition

    1. Required PPE
    2. Technical equipment

376 OSHA and EPA PSM Requirements

OSHA Instruction CPL 2-2.45A

Directorate of Compliance Programs

EXHIBIT 2

DOCUMENT REQUEST LIST

I. PRE-UNIT-SELECTION

    A. OSHA 200 logs for past 3 years

        1. Employer
        2. Contractors

    B. Incident reports

        1. Near miss
        2. Fires
        3. All releases (cross check with EPA documents)

    C. Site plan/Facility overview

    D. Simplified flow diagrams

    E. All permit procedures

        1. Confined space
        2. Hot work
        3. Others

    F. Hazard communication

    G. Overall emergency response plan (emergency action plan, evacuation plan)

    H. Lockout/Tagout

    I. PPE plan/Requirements

    J. Audits

        1. Internal
        2. Corporate
        3. Contracted
        4. Insurance/Consultant

    K. Fire brigade records

        1. Organizational statement
        2. Training records
        3. Callouts/Responses
        4. Roster

Appendix 9 - OSHA Compliance Directive 377

OSHA Instruction CPL 2-2.45A

Directorate of Compliance Programs

        5. Equipment inspection

- L. Respirator program and inspections (emergency use)
- M. Infection/Exposure control program (bloodborne)
- N. Safety and health outline
    1. Minutes of safety and health committee meetings and walkaround reports
    2. Committee roster
- O. Disaster preparedness program
- P. Facility description
    1. Size, capacity, age (units)
    2. History
- Q. Turnaround/Shutdown schedule (not turnaround plan)
- R. Safety and health complaints
- S. Accident investigation logs
- T. Industry hazard alerts (fire and explosion information from other facilities) ("Lessons Learned" by API)
- U. Process hazard analysis scheduling procedure

II. **UNIT-SPECIFIC DOCUMENTS**

- A. Written operating procedures
    1. All current procedures
        a. Normal
        b. Abnormal
        c. Emergency
    2. Startup procedures
        a. Partial (swoop down procedures)
        b. Full (cold)
    3. Shutdown procedures
        a. Normal
        b. Emergency
    4. Upset conditions (beyond normal operating parameters)

378 OSHA and EPA PSM Requirements

OSHA Instruction CPL 2-2.45A

Directorate of Compliance Programs

B. Process safety information

    1. Process chemistry

    2. Capacity (volume)

    3. Operating temperatures and pressures

        • Alarm settings (high, high-high, low, low-low, etc.)

    4. Operating parameters

    5. Consequences of deviations

    6. Flow rates

C. Operating logs (past 6 months)

    1. Foreman

    2. Operator

    3. Manual and Computer

D. Piping and instrumentation diagrams (P&IDS)

    1. Working (unit level) NOTE: Must be current.

    2. Archival

    3. Simplified (detailed, at a later date)

    4. Product

    5. Utility

    6. Fire protection

E. Training records

    1. Operator and supervisory

    2. Training records (summary) for all safety and health programs

        a. Hazard communication
        b. Emergency response
        c. Bloodborne
        d. Respirators and PPE

Appendix 9 - OSHA Compliance Directive 379

OSHA Instruction CPL 2-2.45A

Directorate of Compliance Programs

      e. SCBA
      f. Fire
      g. Others

F. Permits for the units

- Hot work, confined space, etc.

G. Pressure vessel records

1. For at least 20 different vessels, selection based on age, pressure, temperature, toxic chemical involved (corrosive nature, i.e. sulfuric acid), repair history, environmental stress cracking, etc.

2. Inspection records

    a. All previous records
    b. Analysis of defects
    c. Nondestructive testing records
    d. Inspection schedule and frequency
    e. Internal
    f. External
    g. On-stream
    h. Special
    i. U-1 and U-2 records

3. Inspector qualifications

    a. ASNT or equivalent levels (1, 2, or 3)
    b. Roster of inspectors
    c. Training history and documentation

4. Pressure relief valve (PRV) inspection records

5. Selection criteria for PRV's, vessels, etc.

H. Unit plot plan - detailed

I. Instrumentation calibration records

J. Unit emergency response / Action plan

K. Control room blueprint and schematic

L. Work orders

1. Outstanding

380 OSHA and EPA PSM Requirements

OSHA Instruction CPL 2-2.45A

Directorate of Compliance Programs

- 2. Obtain a sample of completed work order
- 3. Written work order procedure
- 4. All safety work orders

M. Environmental sampling records

- 1. Noise
- 2. Air contaminants/Toxins
- 3. Asbestos

N. Product sampling procedures

O. Calibration records for IH sampling equipment

P. Pre-startup review

Q. Rotating equipment inspection records

- 1. Schedule
- 2. Repair records

R. Operator certification

S. Flare system diagram (P&ID)

T. Process hazard analysis (Haz-Op, What-If, etc.)

U. Piping inspection program

- 1. Records/Results
- 2. Schedule
- 3. Inspector qualifications

Appendix 9 - OSHA Compliance Directive 381

OSHA Instruction CPL 2-2.45A

Directorate of Compliance Programs

EXHIBIT 3

INSPECTION EQUIPMENT

I. PERSONAL PROTECTIVE EQUIPMENT

    A. Standard PPE per directive

        1. Safety shoes
        2. Safety glasses with side shields
        3. Hard hat

    B. Site specific PPE

        1. Hearing protection
        2. Respirators with proper filters/cartridges

    C. Flame retardant clothing/coveralls

    D. Emergency escape packs, where necessary

    E. Supplied-air respirators (only trained CSHOs)

    F. Oxygen and combustible meters

II. SAMPLING EQUIPMENT

    A. Hydrogen sulfide dataloggers/dosimeters

    B. Noise dosimeters

    C. Benzene equipment/media

    D. Sulfuric acid/hydrogen fluoride equipment/media

    E. Asbestos media

    F. Other air contaminants

    G. Charging facilities (area and equipment)

    H. TSD sites - specific requirements

III. TECHNICAL EQUIPMENT

    A. Cameras and video cameras

        1. Company policy regarding use

382 OSHA and EPA PSM Requirements

OSHA Instruction CPL 2-2.45A

Directorate of Compliance Programs

    2. Each CSHO/inspection team equipped with a camera
    3. Careful log of each frame (who, when, where, what)

  B. FILM

    1. Each roll should be identified with CSHO, date, and time prior to developing
    2. Each picture identified with CSHO, date, and time
    3. Film must be developed as soon as possible, and identified (who, what, where, when)
    4. Film log must be maintained with roll number, CSHO, date in for developing, date returned
    5. Photos should be mounted on worksheets and identified sequentially for each team member
    6. Negatives must be identified and secured (preferably stored separately from developed photographs)

  C. Videotapes

    1. Identified with CSHO, date, and subject
    2. Original tapes must be maintained
    3. Videotape log maintained with CSHO, camera number (serial number), and date

  D. Audiotapes

    1. Primarily for interviews and/or field notes
    2. Company policy
    3. Permit requirements
    4. Original tapes must be retained in file
    5. Transcription (as needed)
    6. Tapes must be identified with date, team member, and subject matter
    7. Tapes must be logged

INSPECTION KIT

I. Office Supplies

  A. Folders (file folders and expandable)
  B. Paper clips
  C. Hole punch
  D. Stapler and staples
  E. Staple puller
  F. White out/correction tape
  G. Colored pencils/markers
  H. Scissors
  I. Post-its
  J. Tape
  K. Labels

Appendix 9 - OSHA Compliance Directive

OSHA Instruction CPL 2-2.45A

Directorate of Compliance Programs

- L. Pens/pencils
- M. Calculator
- N. Ruler/graph paper
- O. Filing Boxes
- P. Envelopes

II. Inspection Supplies

- A. OSHA forms (1b, photo mounting, 5(a)(1) letters, willful and 5(a)(1) worksheets)
- B. Film, audio and video tapes
- C. Batteries and battery packs for camcorders
- D. Film processing envelopes
- E. Sampling media
  1. smoke tubes
  2. scintillation vials
  3. filters/charcoal tubes
- F. OSHA 31's & travel vouchers

III. Command Center Equipment

- A. Computers (two or more)
  1. database management program
  2. word processor
  3. spreadsheet
  4. floppy disks
- B. Printer with paper and spare ribbon
- C. Disk storage boxes
- D. Fax with extra paper
- E. Cellular phone and pagers if needed
- F. Chargers for all equipment
- G. Answering machine
- H. Telephone directory (OSHA contacts)

384 OSHA and EPA PSM Requirements

OSHA Instruction CPL 2-2.45A

Directorate of Compliance Programs

Library/Reference Material

A. API 510, 750, and others

B. PUB 8-1.5, CPL2 PETROSEP (March 9, 1992)

C. 1910.119 Process Safety Management

D. OTI-PSM (Courses 330/340) manuals

E. 2 sets General Industry and Construction Standards

F. SAVEs manual

G. Field Operations Manual

H. Other references as needed; e.g., NFPA, ANSI, ASNT

Appendix 9 - OSHA Compliance Directive 385

OSHA Instruction CPL 2-2.45A

Directorate of Compliance Programs

Appendix H

Recording PSM-Related Inspection in IMIS

Information about PSM-related inspection activity, as described at H. of this instruction, shall be recorded in IMIS following current instructions in the IMIS manual. These guidelines shall apply:

1. <u>PQV Inspections</u>. The identifier code "PSMPQV" shall be used for these inspections.

    a. PQV inspections, as described at J., K., and L. of this instruction, shall be identified by recording "PSMPQV" in item 25.d of the OSHA-1 Form.

    b. Any inspections of onsite contractors shall also be identified by recording "PSMPQV" in item 25.d of the OSHA-1 Form.

    c. Linkage of all of the employers inspected on-site shall be performed in accordance with the instructions for entering <u>Multi-Employer Inspections</u> currently specified in Chapter V, item E.(5.), of the IMIS Forms Manual.

    d. PQV inspections may be programmed or unprogrammed; all PQV inspections shall be identified as comprehensive.

2. <u>Unprogrammed PSM-related Inspections</u>. All unprogrammed inspection activity relating to the PSM standard--as described at H.3. of this instruction--shall be coded as follows in Item 42, Optional Information of the OSHA-1 form:

Type	ID	Value
N	06	PSMP

    This shall apply to all unprogrammed inspections in which compliance with the PSM standard is investigated; i.e., inspections in which the establishment:

    a. Is not in one of the SIC codes listed in Appendix C of this instruction; or

    b. Is not an establishment selected for a PQV inspection, although it is in one of the SIC codes listed in Appendix C of this instruction.

386 OSHA and EPA PSM Requirements

OSHA Instruction CPL 2-2.45A

Directorate of Compliance Programs

3. **Other Programmed Inspections: Screening for PSM Coverage.** In all programmed safety and health inspections in general industry, a determination shall be made as to whether the establishment is covered by the PSM standard. The establishments shall be coded as follows in Item 42, Optional Information of the OSHA-1 form:

   a. Establishments determined to be covered by the PSM standard:

Type	ID	Value
N	06	PSMY

   b. Establishments determined to be **not** covered by the PSM standard:

Type	ID	Value
N	06	PSMN

REFERENCES

Andrews, D.C. 1993. "Industry Views on Chemical Process Safety." *Process Saf. Progress*, 12(2): 104-105.

Barrish, R.A. 1993. "Process Safety Management in Delaware." *Process Saf. Progress*, 12(2): 115-117.

Burk, A.F. and Smith, W.L. 1990. "Process Safety Management Within Du Pont." *Plant / Oper. Prog.*, 9(4): 269-271.

Carmody, T.W. and Gideaon, J.A. 1992. "Process Safety Management: Resources from the American Institute of Chemical Engineers for Use by Industrial Hygienists." *Amer. Indus. Hygiene Ass'n J.*, 53(6): 404-410.

Center for Chemical Process Safety. 1989. *Guidelines for Technical Management of Process Safety*. AIChE, New York.

Haggin, J. 1990. "Chemical Process Safety Emphasis Shifts to Management." *Chem. Eng. News*, 68(25): 19-27.

Herbert, D.A. 1990. "Process Safety Management Involves Personnel, Facilities and Technology." *Occup. Health & Saf.*, 59(6): 28-30, 32.

Horner, R.A. 1989. "Direction of Plant Process Safety Regulations in the United States." *J. Loss Prev. Process Indus.*, 2(3): 123-125.

Howard, W.B. 1989. "On the Need for Small Chemical Operations to Know and Use Proper Chemical Process Safety Technology." *Plant / Oper. Prog.*, 8(2): 65-69.

Howard, W.B. 1988. "Process Safety Technology and the Responsibility of Industry." *Chem. Eng. Prog.*, 84(9): 25-33.

Kearney, K.E. 1993. "Process Safety Management: An Overview of 1910.119 from a Hazardous Waste Facility Perspective." *Prof. Saf.*, 38(8): 16-22.

Krivan, S.P. 1986. "Avoiding Catastrophic Loss: Technical Safety Audit and Process Safety." *Prof. Saf.*, 31(2): 21-26.

McKelvey, T., Rothschild, M., Gideon, J., Beasley, A., Gressel, M. 1992. "Process Hazards Review Applied to the Use of Anhydrous Ammonia in Agriculture: An Example of Chemical Process Safety for Small Business." *J. Loss Prev. Process Indus.*, 5(5): 297-303.

Moretz, S. 1988. "Process Safety: Controlling Hazardous Chemicals." *Occup. Hazards*, 50(5): 69-72.

Sharkey, J. J., Cutro, R. S., Fraser, W. J., Wildman, G. T. 1992. "Process Safety Testing Programme for Reducing Risks Associated With Large Scale Chemical Manufacturing Operations." *Plant / Oper. Prog.*, 11(4): 238-246.

Sweeney, J.C. 1992. "Measuring Process Safety Management." *Plant / Oper. Prog.*, 11(2): 89-101.

Tomfohrde, J. H. 1985. "Design for Process Safety." *Hydrocarbon Process*, 64(12): 71-74.

Toth, W.C. 1990. "Process Safety Management Implementation of Small Locations." *Plant / Oper. Prog.*, 9(4): 244-245.

ACRONYMS AND ABBREVIATIONS

AIChE	American Institute of Chemical Engineers
ANSI	American National Standards Institute
API	American Petroleum Institute
ARCHIE	Automated Resources for Chemical Hazard Incident Evaluation
ARIP	Accidental Release Information Program
ASTM	American Society for Testing and Materials
CAA	Clean Air Act
CAAA	Clean Air Act Amendments of 1990
CARP	Chemical Accidental Release Prevention
CAS	Chemical Abstract Service
CCPS	Center for Chemical Process Safety (AIChE)
CEPP	Chemical Emergency Preparedness Program
CERCLA	Comprehensive Environmental Response, Compensation, and Liability Act
Chem SEP	Special Emphasis Program for the Chemical Industry
CMA	Chemical Manufacturers Association
CSHO	Compliance Safety and Health Officer
DOT	U.S. Department of Transportation
EHS	Extremely hazardous substance
EPA	U.S. Environmental Protection Agency
EPCRA	Emergency Planning and Community Right-to-Know Act
FMEA	Failure Mode and Effect Analysis
HAZOP	Hazard and Operability Study
HAZWOPER	Hazardous Waste Operations and Emergency Response
JGI	John Gray Institute
LEPC	Local Emergency Planning Committee

390 Acronymns and Abbreviations

MOC	Management of Change
MSDS	Material Safety Data Sheet
NESHAP	National Emissions Standards for Hazardous Air Pollutants
NFPA	National Fire Protection Association
OMB	Office of Management and Budget
ORC	Organization Resources Counselors
OSHA	Occupational Safety and Health Administration
P&ID	Piping and Instrument Diagram
PHA	Process Hazard Analysis
PRIA	Preliminary Regulatory Impact Analysis
PSM	Process Safety Management
PSSR	Pre-Startup Safety Review
RCRA	Resource Conservation and Recovery Act
RFC	Request for Change
RIA	Regulatory Impact Analysis
RMP	Risk Management Plan (or Program)
SARA	Superfund Amendments and Reauthorization Act
SERC	State Emergency Response Committee
SHI	Substance Hazard Index
SIC	Standard Industrial Classification
SOCMA	Synthetic and Organic Chemical Manufacturers Association
SOP	Standard Operating Practice
TCPA	Toxic Catastrophe Prevention Act (New Jersey)
TQ	Threshold Quantity

INDEX

ACCIDENTAL RELEASE
INFORMATION PROGRAM 201
Accidental Release Information Program
(ARIP) 201
American Petroleum Institute 202
Center for Chemical Process Safety 202
Chemical Manufacturers' Association
(CMA) 202
Organization of Economic Cooperation
and Development 201
Responsible Care SuperTM Program 202
Seveso Directive 202

ACCIDENTAL RELEASE
PREVENTION 216
Accident Investigation 229
Accidental Release Prevention
Provisions, Appendix 8 307
Application of PHA Techniques 220
Compiling Process Safety Information
222
Development of a Management System
218
Evaluation of Root-Causes and Near-
Misses 229
Maintenance (Mechanical Integrity) 225
Management of Change 227
Monitoring and Detection Systems 221
Pre-Startup Review 226

Prevention Program Elements 217
Process Hazard Analysis 218
Safety Audits 228
Standard Operating Procedures (SOPs)
223
Training Requirements 224

ACCIDENTAL RELEASE SCENARIOS
199
State Programs 199
Worst-Case Release Scenario 199

BLOCK FLOW DIAGRAM AND
SIMPLIFIED PROCESS FLOW
DIAGRAM
Appendix 3 279

CHEMICAL ACCIDENTAL RELEASE
PREVENTION, See ACCIDENTAL
RELEASE PREVENTION

CHEMICAL INCIDENTS 1
American Petroleum Institute (API) 4
Bhopal 3
Center for Chemical Process Safety 4
Chemical Manufacturers Association
(CMA) 4
Hazard Communication Standard 3

392 Index

Industry Efforts Concerning Process Safety Management 4
International Confederation of Free Trade Unions 4
International Efforts Concerning Process Safety 2
International Federation of Chemical, Energy and General Workers' Unions 4
International Labour Organization 2
Major Incidents 2
Organization of Economic and Cooperative Development 2
Organization Resources Counselors (ORC) 4
Other OSHA Safety Standards 3
PEPCON Plant Oxidizer Accident in Henderson, Nevada 4
Seveso Directive 2
Superfund Amendments and Reauthorization Act (SARA) 2
U.S. Regulation of Hazardous Chemicals 2
United Steelworkers of America 4
World Bank 2

CLEAN AIR ACT AMENDMENTS OF 1990 202
Chemical Safety and Hazard Investigation Board 203

COMPLIANCE AUDITS 185
Audit Team 190
Checklist of Items to Examine During the Audit 192
Compliance Checklist 188
Corrective Action 191
Document Appropriate Response 186
Person Knowledgeable in Audit Techniques 187
Person Knowledgeable in the Process 186
Preparing for the Audit 192
PSM Standard Compliance Guidelines 186
Retention of Two Most Recent Compliance Audit Reports 186
Review of Relevant Documentation 190
Specific Requirements 185
Tracking System 191
Verification Sheet Format 187

CONTRACTORS 114
Compliance Checklist 123
Compliance Guidelines 121
Contract Employer Responsibilities 116
Contractor Training Settlement 128
Contractors, Use of 115
John Gray Institute 114
Nonroutine Work Authorizations 128

COST ESTIMATES 206
Cost Impacts of the PSM Standard 13
Cost of Compliance, EPA Rule 197
Process Hazard Analysis, EPA Rule 206
Registration Costs, EPA Rule 207

DEFINITIONS 28
Atmospheric Tank 28
Boiling Point 28
Catastrophic Release 28
EPA Rule, 210
Facility 29
Highly Hazardous Chemical 29
Hot Work 29
Normally Unoccupied Remote Facility 29
Process 29
Replacement in Kind 29
Trade Secret 29

EMERGENCY PLANNING AND RESPONSE, OSHA RULE 174
Alarm System 175-76
Compliance Checklist 178

Index 393

Compliance Guidelines 175
Drills or Simulated Exercises 175
Emergency Control Center 185
Hazardous Waste Operations and
Emergency Response (HAZWOPER)
Standard 184
Implementation of Emergency Action
Plan 174
OSHA Hazardous Waste and Emergency
Response Regulations 174
Outdoor Processes 176
Plant and Local Community
Coordination of Emergency Response
184
Pre-Planning and Training 175
Pre-Planning for Releases 184
Tertiary Lines of Defense 176
Training 177
Wind Direction Indicator 176

EMERGENCY RESPONSE, EPA
RULE 230
Coordination with LEPC Plans 231
Emergency Response Plan 230
Hazardous Waste Operations and
Emergency Response (HAZWOPER)
231
Written Procedures 231

EMPLOYEE PARTICIPATION 69
Compliance Checklist 71

EMPLOYEE TRAINING PROGRAMS
104
Compliance Checklist 111
Compliance Guidelines 109
Training Examples 109
Training Program Evaluation 110

EPA LIST OF REGULATED
SUBSTANCES AND THRESHOLDS
FOR ACCIDENTAL RELEASE
PREVENTION
Appendix 7 289

EPA RULE, IMPETUS BEHIND 200
Air Toxics Strategy 200
Bhopal, India 200
Chemical Emergency Preparedness
Program (CEPP) 200
Institute, West Virginia 200

EPA RULE, OTHER APPROACHES
CONSIDERED 255

EPA RULE, PENALTIES 240

EPA RULE, PURPOSE OF 208
Affected Areas at Facilities 209
Applicability of the EPA Rule 209
Batch Processors 209
Cost of Compliance 197
Definitions 210
EPA Prevention Program Components
198
Facilities Affected 198, 209
Integrated Management System 208
Management Commitment 208
Petition Process 211
Regulated Substances and Thresholds
210
Risk Management Program Elements
211
Significant Accidental Release 210
Worst-Case Release 210

EPA RULE, SECTION BY SECTION
REVIEW OF 240

EPA'S RULE AND OSHA'S PSM
STANDARD, DIFFERENCES
BETWEEN 248
Compliance Dates 254

394 Index

Contractors, Use of 253

HAZARD ASSESSMENT, EPA RULE 211
Five-Year History of Releases 216
Generic Methodologies for Assessing Off-Site Impacts 214
Other Accidental Release Scenarios 212
Potential Off-site Consequences 213
Range of Events 215
Updating of Off-site Consequences Analyses 215
Worst-Case Release Scenarios 211

HOT WORK PERMIT 30
Compliance Checklist 31

INCIDENT INVESTIGATION 164
Compliance Checklist 171
Compliance Guidelines 167
Five-Year Retention Period 166
Incident investigation Team 165
Incident Involving Contract Employer's Work 165
Investigation Report 165-66
Resolutions and Corrective Actions 165
Sample Investigation Report 167
Specific Requirements 165
System to Report Findings 166

INVESTIGATION AND RESPONSE TO PROCESS INCIDENTS 161
Bhopal 162
Case Histories of Process Incidents 161
Flame Arrester Incident 162
Flixborough, England 161
Inadequate Venting of Exothermic Runaway Reactions 164
Need for Hard-Wired Backup of Critical Safety Instrumentation 163
Seveso, Italy 162

MANAGEMENT OF CHANGE 147
Authorization Requirements 147
Compliance Checklist 150
Compliance Guidelines 148
Developing a Management of Change System 145
Impact of Change on Employee Safety and Health 147
Modifications to Operating Procedures 147
Necessary Time Period for Change 147
Request for Change Form 153
Sample Request for Change Form 154
Technical Basis for Proposed Change 147
Temporary Changes 149

MECHANICAL INTEGRITY OF EQUIPMENT 130
Applicable Codes and Standards 143
Compliance Checklist 136
Compliance Guidelines 134
Constructing New Plants and Equipment 131
Controls 130
Deficiencies in Equipment 134
Emergency Shutdown Systems 130
Equipment Deficiencies 131
Generally Accepted Good Engineering Practices 133
Inspection Methodologies 143
Inspections and Tests 131-33
Lines of Defense 142
Mechanical Integrity Program, Elements of 135
Piping Systems 130
Pressure Vessels and Storage Tanks 130
Pumps 130
Quality Assurance of Mechanical Equipment 134
Quality Assurance System 144
Relief and Vent Systems 130

Index   395

Training 132
Written Procedures 132

OPERATING PROCEDURES 74
Compliance Checklist 79
Computerized Process Control Systems 77
Operating Procedures and Practices 77
Standard Operating Practices (SOPs) 78
Training for Handling Upset Conditions 78

OSHA AND EPA RULES, COMPARISON OF 198
Threshold Quantities 199

OSHA COMPLIANCE DIRECTIVE 46
CPL 2-2.45A, Appendix 9 330

OSHA INSPECTIONS AND ENFORCEMENT 46
Inspection Totals 47
PSM Provisions Cited 48
Settlement of 1989 Phillips Petroleum Explosion 51
Specific Violations 49
Violations of Operating Procedures 48

OSHA LIST OF HIGHLY HAZARDOUS CHEMICALS, TOXICS, AND REACTIVES
Appendix 2 275

PRE-STARTUP SAFETY REVIEWS 155
Compliance Checklist 158
Compliance Guidelines 157
PSM Standard Requirements 155

PROCESS HAZARD ANALYSIS, BENEFITS FOR SMALL BUSINESSES 77

PROCESS HAZARD ANALYSIS, COMPLIANCE GUIDELINES 93
Application of PHA to Specific Processes 101
Batch Type Processes 101
Compliance Checklist 94
Gas Plant 102
PHA Team 101
Selection of PHA Methodology 93
Small Businesses 102
Standard Boiler or Heat Exchanger 101

PROCESS HAZARD ANALYSIS, METHODOLOGIES 83
Checklist 84
Failure Mode and Effect Analysis (FMEA) 83, 85-86
Fault Tree Analysis 86
Hazard and Operability Study (HAZOP) 83, 85-86
What-If 84
What-If/Checklist 84, 86

PROCESS HAZARD ANALYSIS, SPECIFIC REQUIREMENTS 88

PROCESS HAZARD ANALYSIS, TEAM APPROACH 91

PROCESS SAFETY INFORMATION 59
American Institute of Chemical Engineers 63
American National Standards Institute 63
American Petroleum Institute 63
American Society for Testing and Materials 63
American Society of Exchange Manufacturers Association 63

396 Index

American Society of Mechanical Engineers 63
Block Flow Diagram 61-62
Compliance Checklist 65
Compliance Tips 62
Information on Equipment 60
Information on Hazards 59
Information on Technology 60
Material Safety Data Sheet (MSDS) 61-62
National Association of Corrosion Engineers 63
National Board of Boiler and Pressure Vessel Inspectors 63
National Fire Protection Association 63
Piping and Instrument Diagrams (P&IDs) 62-63
Simplified Process Flow Diagram 61

PROCESS SAFETY OPERATING PROCEDURES 73
Operating Limits 74
Safety and Health Considerations 74
Steps for Each Operating Phase 73
Written Operating Procedures 73

PSM STANDARD, APPLICATION OF 20
American Petroleum Institute 25
American Petroleum Institute's RP 750 22
Batch Processing Operations 27
Chemical Manufacturers Association (CMA) 27
Delaware's Extremely Hazardous Substances Risk Management Act 22
EHS list 25
EPCRA and CAAA Chemical Lists, Role of 24
Exemptions for Hydrocarbon Fuels 25
Flammable Liquids Exemption 25
List of Highly Hazardous Chemicals 22
Meaning of the Term "Processes" 21
Minimum Threshold Levels 23
National Fire Protection Association's NFPA 49 22
New Jersey's Toxic Catastrophe Prevention Act 22
Organization Resources Counselors 22, 25
Other Exemptions 26
Seveso Directive 22
Synthetic and Organic Chemical Manufacturers Association (SOCMA) 27
Turner-Described Gaussian Dispersion Model 23

PSM STANDARD, BASIC REQUIREMENTS 19
Bhopal 18
Block Flow Diagram and Simplified Process Flow Diagram, Appendix 3 279
Chem-SEP 19
Compliance Deadlines 20
Hazard Communication Standard 17
Impetus Behind PSM Standard 18
Major Accidents 18
Performance-Oriented Approach 18
"Process" 19
Process Hazard Analysis (PHA) 19
PSM Standard (29 CFR 1910.119), Appendix 1 261
Sources of Further Information, Appendix 4 281

PSM STANDARD, DEVELOPMENT OF 5
Clean Air Act Amendments of 1990, Impact of 6
Institute, West Virginia 5
Organization Resources Counselors (ORC) 6
Special Emphasis Program for the Chemical Industry (Chem SEP) 6

Index 397

PSM STANDARD, EFFECTIVENESS
RATE OF 13
Control of Hazardous Energy Source
(Lockout/Tagout) 13
Cost Impacts of the PSM Standard 13
Electrical Safety-Related Work
Practices 13
Impact on Small Businesses 14
Kearney/Centaur Report 13
Permit Required Confined Spaces 13
Preliminary Regulatory Impact Analysis
(PRIA) 13
Sunset Provision Unnecessary 15

PSM STANDARD, INDUSTRIES
AFFECTED 9
Concerns Raised by the Office of
Management and Budget 11
Concerns Raised During the Rulemaking
Process 9
Consideration of Other Regulatory
Options 11
Organization Resources Counselors 10
Standard Industrial Classification (SIC)
9

PSM SYSTEM, COMPONENTS OF 53
Accountability 55
Audits and Corrective Actions 57
Capital Project Review and Design
Procedures 55
Compiling Process Safety Information
59
Controlling 54
Enhancement of Process Safety
Knowledge 57
Generic Characteristics of a
Management System 54
Human Factors 56
Implementing 54
Incident Investigation 56
Management of Change 56

Objective and Goals 55
Organizing 54
Planning 54
Process and Equipment Integrity 56
Process Knowledge and Documentation
55
Process Risk Management 55
Standards, Codes, and Laws 57
Training and Performance 56

REGULATED CHEMICALS AND
THRESHOLD QUANTITIES 12
EPA List of Regulated Substances and
Thresholds for Accidental Release
Prevention, Appendix 7 289
OSHA List of Highly Hazardous
Chemicals, Toxics, and Reactives,
Appendix 2 275

RISK MANAGEMENT PLAN AND
DOCUMENTATION 232
Auditing of RMPs 236
Chemical Safety and Hazard
Investigation Board 232
Implementation of Registration
Requirements 239
Purpose of the RMP 232
Registration: Information Required 237
RMP Content 233
Submission of RMPs 235

RISK MANAGEMENT PROGRAMS,
CLEAN AIR ACT REQUIREMENTS
Hazard Assessment 204
Chemical Safety and Hazard
Investigation Board 204-05
Emergency Response Program 205
Hazard Assessment 205
Prevention Program 205
Registration 204

STATE LAWS, ROLE OF 43, 254
Accidental Release Scenarios 199
California 45
Delaware 43
Directory of Consultation
Programs,Appendix 5 284
States With Approved OSHA Programs,
Appendix 6 286

TRADE SECRETS 38
Compliance Checklist 41
Compliance Safety and Health Officer
(CSHO) 38
Freedom of Information Act 38
Hazard Communication Standard 39

TRAINING 115
Application 106
Clean Air Act Training Requirements
108
Grandfathering 106
Refresher Training 107
Training Documentation 107